Strong Medicine— Chemistry at the Pharmacy

Science in Our World
Volume Two

Developed in collaboration with

Hoechst Marion Roussel, Inc.

Series Editor
Mickey Sarquis, Director
Center for Chemical Education

©1995 by Terrific Science Press
ISBN 1-883822-10-6 First printed 1995 Revised and reprinted 1997
All rights reserved. Printed in the United States of America.

This project was supported, in part, by the National Science Foundation. Any opinions, findings, and conclusions or recommendations expressed in this material are those of the authors and do not necessarily reflect the views of the National Science Foundation. The Government has certain rights to this material. This material is based upon work supported by the National Science Foundation under Grant No. TPE-9153930.

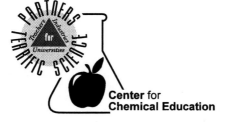

Center for
Chemical Education

This monograph is intended for use by teachers, chemists, and properly supervised students. Teachers and other users must develop and follow procedures for the safe handling, use, and disposal of chemicals in accordance with local and state regulations and requirements. The cautions, warnings, and safety reminders associated with the doing of experiments and activities involving the use of chemicals and equipment contained in this publication have been compiled from sources believed to be reliable and to represent the best opinion on the subject as of 1995. However, no warranty, guarantee, or representation is made by the editor, contributors, Hoechst Marion Roussel, Inc., or the Terrific Science Press as to the correctness or sufficiency of any information herein. Neither the editor, contributors, Hoechst Marion Roussel, nor the publisher assumes any responsibility or liability for the use of the information herein, nor can it be assumed that all necessary warnings and precautionary measures are contained in this publication. Other or additional information or measures may be required or desirable because of particular or exceptional conditions or circumstances, or because of new or changed legislation.

Contributors

Industrial Mentor

Roger Parker
Senior Research Chemist, Hoechst Marion Roussel, Inc.
Cincinnati, Ohio

Academic Mentor

Dan McLoughlin
Chemistry Department, Xavier University
Cincinnati, Ohio
Partners for Terrific Science Project Co-Director

Peer Mentor

Ginger Tannenbaum
Chemistry Teacher, Fairfield High School
Fairfield, Ohio

Principal Investigators

Mickey Sarquis	Miami University, Middletown, Ohio
Jim Coats	Dow Chemical USA (retired), Findlay, Ohio
Dan McLoughlin	Xavier University, Cincinnati, Ohio
Rex Bucheit	Fillmore Elementary School, Hamilton, Ohio

Partners for Terrific Science Advisory Board

Ruby L. Bryant	Colonel White High School, Dayton, Ohio
Rex Bucheit	Fillmore Elementary School (ex-officio), Hamilton, Ohio
Jim Coats	Dow Chemical USA (retired, ex-officio), Findlay, Ohio
Dick French	Quantum Chemical Corporation (retired, ex-officio), Cincinnati, Ohio
Judy Gilbert	BP America/Ohio Chemical Council, Lima, Ohio
Linda Jester	John XXIII Elementary School, Middletown, Ohio
James C. Letton	Procter & Gamble, Cincinnati, Ohio
Ted J. Logan	Procter & Gamble, Ross, Ohio
Ken Lohr	Hoechst Marion Roussel, Inc. (retired), Cincinnati, Ohio
Alan McClelland	Delaware Science Alliance (DuPont, retired), Rockland, Delaware (deceased)
Dan McLoughlin	Xavier University (ex-officio), Cincinnati, Ohio
Raymond C. Odioso	R.C. Odioso Consultants, Inc. (Drackett, retired), Cincinnati, Ohio, St. Petersburg Beach, Florida
Tom Runyan	Garfield Alternative School, Middletown, Ohio
Ken Wilkinson	Hilton Davis Company (retired), Cincinnati, Ohio
John P. Williams	Miami University Hamilton, Hamilton, Ohio
Regina Wolterman	Our Lady of Lourdes Elementary School, Cincinnati, Ohio

Table of Contents

Acknowledgments

The authors and editor wish to thank the following individuals who have contributed to the development of the *Science in Our World* series of Teacher Resource Modules.

Terrific Science Press Design and Production Team
Susan Gertz, Amy Stander, Lisa Taylor, Thomas Nackid, Stephen Gentle, Vickie Fultz, Anne Munson, Amy Hudepohl, Andrea Nolan, Pamela Mason

Reviewers

Frank Cardulla	Niles North High School, Skokie, Illinois
Susan Hershberger	Miami University, Oxford, Ohio
Baird Lloyd	Miami University, Middletown, Ohio
Mark Sabo	Miami University, Middletown, Ohio
Dave Tomlin	Wright Patterson Air Force Base, Dayton, Ohio
Linda Woodward	University of Southwestern Louisiana, Lafayette, Louisiana

Center for Chemical Education Staff

Mickey Sarquis, Director
Bruce L. Peters, Jr., Associate Director
Billie Gerzema, Administrative Assistant

Assistants to Director

Susan Gertz	Mark Sabo
Lynn Hogue	Lisa Meeder Turnbull

Project Coordinators and Managers

Richard French	Andrea Nolan
Betty Kibbey	Ginger Smith
Carl Morgan	Amy Stander

Research Associates and Assistants

Kersti Cox	Pamela Mason
Stephen Gentle	Anne Munson
Susan Hershberger	Thomas Nackid
Amy Hudepohl	Michael Parks
Robert Hunter	Lisa Taylor

Program Secretaries

Victoria Burton	Ruth Willis

Graduate Assistants

Michelle Diebolt	Richard Rischling
Nancy Grim	Michella Stultz

Foreword

Strong Medicine—Chemistry at the Pharmacy is one of the *Science in Our World* Teacher Resource Modules. This set is aimed at enabling teachers to introduce their students to the concepts and processes of industrial chemistry and relate these concepts to the consumer products students encounter daily. These hands-on, problem-solving activities help connect science lessons with real life.

Developed as a collaborative effort between industrial, academic, and teacher peer mentors in the *Partners for Terrific Science* program, this module provides background information on the pharmaceutical industry and Hoechst Marion Roussel's role in this industry, as well as a content review of pharmaceutical chemistry and pedagogical strategies. The activities in this module have been tested by participants in *Partners* programs and by *Partners* teachers in their classrooms, and reviewed by experts in the field to help ensure accuracy, safety, and pedagogical effectiveness.

Partners for Terrific Science, established in 1986, is an industrial/academic partnership that facilitates interaction among classroom teachers, industrial scientists and engineers, and university chemistry faculty to make science education more interesting, relevant, and understandable for all students. The partnership is supported by the Ohio Chemical Council and its more than 100 members, the National Science Foundation, the U.S. Department of Education, the Ohio Board of Regents, the American Chemical Society—Cincinnati Section, Miami University, and over 50 private-sector partners. Hoechst Marion Roussel has generously contributed to the production of this module.

The Teacher Resource Modules have been developed especially for teachers who want to use industry-based physical science activities in the classroom, but who may not have been able to attend a *Partners* workshop at the Miami site or one of the Affiliate sites nationwide. We want to thank all the contributors, participants, and mentors who made this publication possible.

We hope you will find that these Teacher Resource Modules provide you with a useful and exciting way to involve your students in doing chemistry through integrated real-world themes. We welcome your comments at any time and are interested in learning about especially successful uses of these materials.

Mickey Sarquis, Director
Center for Chemical Education
July 1995

The Center for Chemical Education

Built on a tradition of quality programming, materials development, and networking between academia and industry, Miami University's Center for Chemical Education (CCE) encompasses a multifaceted collaboration of cross-grade-level and interdisciplinary initiatives begun in the mid-1980s as Terrific Science Programs. These initiatives are linked through the centrality of chemistry to the goal of fostering quality hands-on, minds-on science education for all students. CCE activities include credit coursework and other opportunities for educators at all levels; K–12 student programs; undergraduate, graduate, and postgraduate programs in chemical education; materials development, including teacher resource materials, program handbooks, and videos; public outreach efforts and networking to foster new and existing partnerships among classroom teachers, university-based science educators, industrial scientists, and professional societies.

Professional Development for Educators

Credit Courses

The Center for Chemical Education offers a variety of summer and academic-year workshop-style courses for K–12 and college teachers. While each workshop has a unique focus, all reflect current pedagogical approaches in science education, cutting-edge academic and industrial research topics, and classroom applications for teachers and students. Short courses provide opportunities for educators to enrich their science teaching in a limited amount of time. All courses offer graduate credit.

Non-Credit Courses

Academies allow CCE graduates and other teachers to attend special one-day sessions presented by leading science educators from around the United States. Offerings include seminars, mini-workshops, and share-and-swap sessions.

Internships

Through 8- to 10-week summer internships, program graduates work as members of industrial teams to gain insight into the day-to-day workings of industrial laboratories, enabling them to bring real-world perspectives into the classroom.

Fellowships

Master teachers at primary, secondary, and college levels do research in chemical education and undertake curriculum and materials development as Teacher Fellows with the Center for Chemical Education. Fellowships are available for the summer and the academic year.

K–12 Student Programming

Summer Camps

A variety of summer camps are available to area elementary, middle, and high school students. These camps not only provide laboratory-based enrichment for students, but also enable educators in summer courses to apply their knowledge of hands-on exploration and leadership skills. Satellite camps are offered at affiliated sites throughout the country.

Science Carnivals

Carnivals challenge elementary school students with hands-on science in a non-traditional atmosphere, encouraging them to apply the scientific method to activities that demonstrate scientific principles. Sponsoring teachers and their students host these carnivals for other students in their districts.

Super Saturday Science Sessions	High school students are introduced to industrial and research applications of science and technology through special Saturday sessions that involve the students in experiment-based problem-solving. Topics have included waste management, environmental sampling, engineering technology, paper science, chemical analysis, microbiology, and many others.
Ambassador Program	Professional chemists, technicians, and engineers, practicing and recently retired, play important roles as classroom ambassadors for high school and two-year college students. Ambassadors not only serve as classroom resources, but they are also available as consultants when a laboratory scenario calls for outside expertise; they mentor special projects both in and out of the classroom; and they are available for career counseling and professional advice.

Undergraduate and Graduate Student Programming

Teaching Science with TOYS Undergraduate Course	This undergraduate course replicates the Teaching Science with TOYS teacher inservice program for the preservice audience. Students participate in hands-on physics and chemistry sessions.
General Chemistry Initiative	This effort is aimed at more effectively including chemical analysis and problem solving in the two-year college curriculum. To accomplish this goal, we are developing and testing discovery-based laboratory scenarios and take-home lecture supplements that illustrate topics in chemistry through activities beyond the classroom. In addition to demonstrating general chemistry concepts, these activities also involve students in critical-thinking and group problem-solving skills used by professional chemists in industry and academia.
Chemical Technology Curriculum Development	Curriculum and materials development efforts highlight the collaboration between college and high school faculty and industrial partners. These efforts will lead to the dissemination of a series of activity-based monographs, including detailed instructions for discovery-based investigations that challenge students to apply principles of chemical technology, chemical analysis, and Good Laboratory Practices in solving problems that confront practicing chemical technicians in the workplace.
Other Undergraduate Activities	The CCE has offered short courses/seminars for undergraduates that are similar in focus and pedagogy to CCE teacher/faculty enhancement programming. In addition, CCE staff members provide Miami University students with opportunities to interact in area schools through public outreach efforts and to undertake independent study projects in chemical education.
Degree Program	Miami's Department of Chemistry offers both a Ph.D. and M.S. in Chemical Education for graduate students who are interested in becoming teachers of chemistry in situations where a comprehensive knowledge of advanced chemical concepts is required and where acceptable scholarly activity can include the pursuit of chemical education research.

Educational Materials

The Terrific Science Press publications have emerged from CCE's work with classroom teachers of grades K–12 and college in graduate-credit, workshop-style inservice courses. Before being released, our materials undergo extensive classroom testing by teachers working with students at the targeted grade level, peer review by experts in the field for accuracy and safety, and editing by a staff of technical writers for clear, accurate, and consistent materials lists and procedures. The following is a list of Terrific Science Press publications to date.

Science Activities for Elementary Classrooms (1986)

Science SHARE is a resource for busy K–6 teachers to enable them to use hands-on science activities in their classrooms. The activities included use common, everyday materials and complement or supplement any existing science curriculum. This book was published in collaboration with Flinn Scientific, Inc.

Polymers All Around You! (1992)

This monograph focuses on the uses of polymer chemistry in the classroom. It includes several multi-part activities dealing with topics such as polymer recycling and polymers and polarized light. This monograph was published in collaboration with POLYED, a joint education committee of two divisions of the American Chemical Society: the Division of Polymer Chemistry and the Division of the Polymeric Materials: Science and Engineering.

Fun with Chemistry Volume 2 (1993)

The second volume of a set of two hands-on activity collections, this book contains classroom-tested science activities that enhance teaching, are fun to do, and help make science relevant to young students. This book was published in collaboration with the Institute for Chemical Education (ICE), University of Wisconsin-Madison.

Santa's Scientific Christmas (1993)

In this school play for elementary students, Santa's elves teach him the science behind his toys. The book and accompanying video provide step-by-step instructions for presenting the play. The book also contains eight fun, hands-on science activities to do in the classroom.

Teaching Chemistry with TOYS Teaching Physics with TOYS (1995)

Each volume contains more than 40 activities for grades K–9. Both were developed in collaboration with and tested by classroom teachers from around the country. These volumes were published in collaboration with McGraw-Hill, Inc.

Palette of Color Monograph Series (1995)

The three monographs in this series present the chemistry behind dye colors and show how this chemistry is applied in "real-world" settings:
- The Chemistry of Vat Dyes
- The Chemistry of Natural Dyes
- The Chemistry of Food Dyes

Science in Our World Teacher Resource Modules (1995)

Each volume of this five-volume set presents chemistry activities based on a specific industry—everything from pharmaceuticals to polymers. Developed as a result of the *Partners for Terrific Science* program, this set explores the following topics and industries:
- Science Fare—Chemistry at the Table (Procter & Gamble)
- Strong Medicine—Chemistry at the Pharmacy (Hoechst Marion Roussel, Inc.)
- Dirt Alert—The Chemistry of Cleaning (Diversey Corporation)
- Fat Chance—The Chemistry of Lipids (Henkel Corporation, Emery Group)
- Chain Gang—The Chemistry of Polymers (Quantum Chemical Corporation)

Teaching Physical Science through Children's Literature (1996)

This book offers 20 complete lessons for teaching hands-on, discovery-oriented physical science in the elementary classroom using children's fiction and nonfiction books as an integral part of that instruction. Each lesson in this book is a tightly integrated learning episode with a clearly defined science content objective supported and enriched by all facets of the lesson, including reading of both fiction and nonfiction, writing, and, where appropriate, mathematics. Along with the science content objectives, many process objectives are woven into every lesson.

Teaching Science with TOYS Teacher Resource Modules (1996, 1997)

The modules in this series are designed as instructional units focusing on a given theme or content area in chemistry or physics. Built around a collection of grade-level-appropriate TOYS activities, each Teacher Resource Module also includes a content review and pedagogical strategies section. Volumes listed below were published or are forthcoming in collaboration with McGraw-Hill, Inc.

- Exploring Matter with TOYS: Using and Understanding the Senses
- Investigating Solids, Liquids, and Gases with TOYS: States of Matter and Changes of State
- Transforming Energy with TOYS: Mechanical Energy and Energy Conversions

Terrific Science Network

Affiliates

College and district affiliates to CCE programs disseminate ideas and programming throughout the United States. Program affiliates offer support for local teachers, including workshops, resource/symposium sessions, and inservices; science camps; and college courses.

Industrial Partners

We collaborate directly with over 40 industrial partners, all of whom are fully dedicated to enhancing the quality of science education for teachers and students in their communities and beyond. A list of corporations and organizations that support *Partners for Terrific Science* is included on the following page.

Outreach

On the average, graduates of CCE professional development programs report reaching about 40 other teachers through district inservices and other outreach efforts they undertake. Additionally, graduates, especially those in facilitator programs, institute their own local student programs. CCE staff also undertake significant outreach through collaboration with local schools, service organizations, professional societies, and museums.

Newsletters

CCE newsletters provide a vehicle for network communication between program graduates, members of industry, and other individuals active in chemical and science education. Newsletters contain program information, hands-on science activities, teacher resources, and ideas on how to integrate hands-on science into the curriculum.

For more information about any of the CCE initiatives, contact us at

Center for Chemical Education
4200 East University Blvd.
Middletown, OH 45042
513/727-3318
FAX: 513/727-3223
e-mail: *CCE@muohio.edu*
http://www.muohio.edu/~ccecwis/

Partnership Network

We appreciate the dedication and contributions of the following corporations and organizations, who together make *Partners for Terrific Science* a true partnership for the betterment of chemical education for all teachers and students.

Partners in the Private Sector

A & B Foundry, Inc.
Aeronca, Inc.
Ag Renu
Air Products and Chemicals, Inc.
Armco, Inc.
Armco Research and Technology
ARW Polywood
Ashland Chemical Company
Bank One
BASF
Bay West Paper Corporation
Black Clawson Company
BP America: BP Oil, BP Chemicals
Coats & Clark
Crystal Tissue Company
DataChem Laboratories, Inc.
Diversey Corporation
Ronald T. Dodge Company
Dover Chemical Corporation
Dow Chemical USA
Fluor Daniel Fernald, Inc.
Formica
Henkel Corporation, Emery Group

Hewlett-Packard Company
Hilton Davis Company
Hoechst Marion Roussel, Inc.
Inland Container Corporation
Jefferson Smurfit Corporation
JLJ, Inc.
Magnode Corporation
Middletown Paperboard Corporation
Middletown Regional Hospital
Middletown Wastewater Treatment Plant
Middletown Water Treatment Plant
Miller Brewing Company
The Monsanto Fund
Owens Corning Science & Technology Laboratories
The Procter & Gamble Company
Quality Chemicals
Quantum Chemical Corporation
Rumpke Waste Removal/Recycling
Shepherd Chemical Company
Shepherd Color Company
Sorg Paper Company
Square D Company
Sun Chemical Corporation

Partners in the Public Sector

Hamilton County Board of Education
Indiana Tech-Prep
Miami University
Middletown Clean Community
National Institute of Environmental Health Sciences
National Science Foundation
Ohio Board of Regents, Columbus, OH

Ohio Department of Education
Ohio Environmental Protection Agency
Ohio Tech-Prep
State Board for Technical and Comprehensive Education, Columbia, SC
US Department of Education
US Department of Energy, Cincinnati, OH

Professional Societies

American Association of Physics Teachers
African American Math-Science Coalition
American Chemical Society— Central Regional Council
American Chemical Society— Cincinnati Section
American Chemical Society— Dayton Section
American Chemical Society—POLYED
American Chemical Society— Technician Division
American Chemical Society, Washington, DC

American Institute of Chemical Engineers
Chemical Manufacturers Association
Chemistry Teachers Club of New York
Intersocietal Polymer and Plastics Education Initiative
Minorities in Mathematics, Science and Engineering
National Organization of Black Chemists and Chemical Engineers—Cincinnati Section
National Science Teachers Association
Ohio Chemical Council
Science Education Council of Ohio
Society of Plastics Engineers

More than 3,000 teachers are involved in and actively benefiting from this Network.

An Invitation to Industrial Chemists

It is not unusual to hear children say they want to be doctors, astronauts, or teachers when they grow up. It is easy for children to see adults they admire doing these jobs in books, on television, and in real life. But where are our aspiring chemists? The chemist portrayed on television often bears close resemblance to Mr. Hyde: an unrealistic and unfortunate role model.

Children delight in learning and enjoy using words like "stegosaurus" and "pterodactyl." Wouldn't it be wonderful to hear words like "chromatography" and "density" used with the same excitement? You could be introducing elementary school students to these words for the first time. And imagine a 10-year-old child coming home from school and announcing, "When I grow up, I want to be a chemist!" You can be the one responsible for such enthusiasm. By taking the time to visit and interact with an elementary or middle school classroom as a guest scientist, you can become the chemist who makes the difference.

You are probably aware that many non-chemists, including many prehigh school teachers, find science in general (and chemistry in particular) mysterious and threatening. When given a chance, both teachers and students can enjoy transforming the classroom into a laboratory and exploring like real scientists. Consider being the catalyst for this transformation.

Unlike magicians, scientists attempt to find explanations for why and how things happen. Challenge students to join in on the fun of searching for explanations. At the introductory level, it is far more important to provide non-threatening opportunities for the students to postulate "why?" than it is for their responses to be absolutely complete. If the accepted explanation is too complex to discuss, maybe the emphasis of the presentation is wrong. For example, discussions focusing on the fact that a color change can be an indication of a chemical reaction may be more useful than a detailed explanation of the reaction mechanisms involved.

Because science involves the process of discovery, it is equally important to let the students know that not all the answers are known and that they too can make a difference. Teachers should be made to feel that responses like "I don't know. What do you think?" or "Let's find out together," are acceptable. It is also important to point out that not everyone's results will be the same. Reinforce the idea that a student's results are not wrong just because they are different from a classmate's results.

While using the term "chemistry," try relating the topics to real-life experiences and integrating topics into non-science areas. After all, chemistry is all around us, not just in the chemistry lab.

When interacting with students, take care to involve all of them. It is very worthwhile to spend time talking informally with small groups or individual students before, during, or after your presentation. It is important to convey the message that chemistry is for all who are willing to apply themselves to the questions before them. Chemistry is neither sexist, racist, nor frightening.

For more information on becoming involved in the classroom and a practical and valuable discussion of some do's and don'ts, a resource is available. The American Chemical Society Education Division has an informative booklet and video called *Chemists in the Classroom*. You may request this package for $20.00 from: Education Division, American Chemical Society, 1155 Sixteenth Street NW, Washington, DC 20036, 800/227-5558.

How to Use This Teacher Resource Module

This section is an introduction to the Teacher Resource Module and its organization. The industry featured in this module is the pharmaceutical industry.

How Is This Resource Module Organized?

The Teacher Resource Module is organized into the following main sections: How to Use This Teacher Resource Module (this section), Background for Teachers, Using the Activities in the Classroom, and Activities and Demonstrations. Background for Teachers includes Overview of the Pharmaceutical Chemistry Industry, The Hoechst Marion Roussel Research Institute, and Content Review. Using the Activities in the Classroom includes Pedagogical Strategies, an Annotated List of Activities and Demonstrations, and a Curriculum Placement Guide. The following paragraphs provide a brief overview of the *Strong Medicine—Chemistry at the Pharmacy* module.

Background for Teachers

Overviews of the pharmaceutical industry and Hoechst Marion Roussel's role in the industry provide information on the industrial aspect of these activities. The Content Review section is intended to provide you, the teacher, with an introduction to (or a review of) the concepts covered in the module. The material in this section (and in the individual activity explanations) intentionally gives you information at a level beyond what you will present to your students. You can then evaluate how to adjust the content presentation for your own students.

The Content Review section in this module covers the following topics:
- Polymers as Biological Materials
- Chromatography
- Aspirin Products

Using the Activities in the Classroom

The Pedagogical Strategies section provides ideas for effectively teaching a unit on the pharmaceutical industry. It suggests a variety of ways to incorporate the industry-based activities presented in the module into your curriculum. The Annotated List of Activities and Demonstrations and the Curriculum Placement Guide provide recommended grade levels, descriptions of the activities, and recommended placement of the activities within a typical curriculum.

Module Activities

Each module activity includes complete instructions for conducting the activity in your classroom. These activities have been classroom-tested by teachers like yourself and have been demonstrated to be practical, safe, and effective in the typical classroom. The following information is provided for each activity:

Recommended Grade Level: The grade levels at which the activity will be most effective are listed.

Group Size: The optimal student group size is listed.

Time for Preparation:	This includes time to set up for the activity before beginning with the students.
Time for Procedure:	An estimated time for conducting the activity is listed. This time estimate is based on feedback from classroom testing, but your time may vary depending on your classroom and teaching style.
Materials:	Materials are listed for each part of the activity, divided into amounts per class, per group, and per student.
Resources:	Sources for difficult-to-find materials are listed.
Safety and Disposal:	Special safety and/or disposal procedures are listed if required.
Getting Ready:	Information is provided in Getting Ready when preparation is needed prior to beginning the activity with the students.
Opening Strategy:	A strategy for introducing the topic to be covered and for gaining the students' interest is suggested.
Procedure:	The steps in the Procedure are directed toward you, the teacher, and include cautions and suggestions where appropriate.
Variations and Extensions:	Variations are alternative methods for doing the Procedure. Extensions are methods for furthering student understanding.
Discussion:	Possible questions for students are provided.
Explanation:	The Explanation is written to you, the teacher, and is intended to be modified for students.
Key Science Concepts:	Targeted key science topics are listed.
Cross-Curricular Integration:	Cross-Curricular Integration provides suggestions for integrating the science activity with other areas of the curriculum.
References:	References used to write this activity are listed.

Notes and safety cautions are included in activities as needed and are indicated by the following icons and type style:

 Notes are preceded by an arrow.

Cautions are preceded by an exclamation point.

Employing Appropriate Safety Procedures

Experiments, demonstrations, and hands-on activities add relevance, fun, and excitement to science education at any level. However, even the simplest activity can become dangerous when the proper safety precautions are ignored or when the activity is done incorrectly or performed by students without proper supervision. While the activities in this book include cautions, warnings, and safety reminders from sources believed to be reliable, and while the text has been extensively reviewed, it is your responsibility to develop and follow procedures for the safe execution of any activity you choose to do and for the safe handling, use, and disposal of chemicals in accordance with local and state regulations and requirements.

Safety First

- Collect and read the Materials Safety Data Sheets (MSDS) for all of the chemicals used in your experiments. MSDS's provide physical property data, toxicity information, and handling and disposal specifications for chemicals. They can be obtained upon request from manufacturers and distributors of these chemicals. In fact, MSDS's are often shipped with chemicals when they are ordered. These should be collected and made available to students, faculty, or parents for information about specific chemicals in these activities.

- Read and follow the American Chemical Society Minimum Safety Guidelines for Chemical Demonstrations on the next page. Remember that you are a role model for your students—your attention to safety will help them develop good safety habits while assuring that everyone has fun with these activities.

- Read each activity carefully and observe all safety precautions and disposal procedures. Determine and follow all local and state regulations and requirements.

- Never attempt an activity if you are unfamiliar or uncomfortable with the procedures or materials involved. Consult a high school or college chemistry teacher or an industrial chemist for advice or ask him or her to perform the activity for your class. These people are often delighted to help.

- Always practice activities yourself before using them with your class. This is the only way to become thoroughly familiar with an activity, and familiarity will help prevent potentially hazardous (or merely embarrassing) mishaps. In addition, you may find variations that will make the activity more meaningful to your students.

- Undertake activities only at the recommended grade levels and only with adult supervision.

- You, your assistants, and any students participating in the preparation for or doing of the activity must wear safety goggles if indicated in the activity and at any other time you deem necessary.

- Special safety instructions are not given for everyday classroom materials being used in a typical manner. Use common sense when working with hot, sharp, or breakable objects. Keep tables or desks covered to avoid stains. Keep spills cleaned up to avoid falls.

- When an activity requires students to smell a substance, instruct them to smell the substance as follows: hold its container approximately 6 inches from the nose and, using the free hand, gently waft the air above the open container toward the nose. Never smell an unknown substance by placing it directly under the nose. (See figure.)

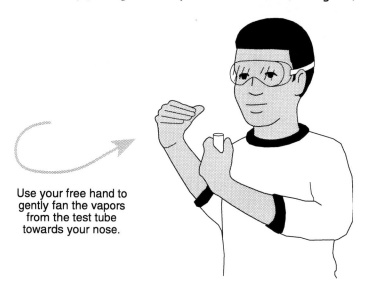

Use your free hand to gently fan the vapors from the test tube towards your nose.

Wafting procedure—Carefully wave the air above the open container towards your nose. Avoid hitting the container in the process.

- Caution students never to taste anything made in the laboratory and not to place their fingers in their mouths after handling laboratory chemicals.

ACS Minimum Safety Guidelines for Chemical Demonstrations

This section outlines safety procedures that Chemical Demonstrators must follow at all times.

1. Know the properties of the chemicals and the chemical reactions involved in all demonstrations presented.

2. Comply with all local rules and regulations.

3. Wear appropriate eye protection for all chemical demonstrations.

4. Warn the members of the audience to cover their ears whenever a loud noise is anticipated.

5. Plan the demonstration so that harmful quantities of noxious gases (e.g., NO_2, SO_2, H_2S) do not enter the local air supply.

6. Provide safety shield protection wherever there is the slightest possibility that a container, its fragments or its contents could be propelled with sufficient force to cause personal injury.

7. Arrange to have a fire extinguisher at hand whenever the slightest possibility for fire exists.

8. Do not taste or encourage spectators to taste any non-food substance.

9. Never use demonstrations in which parts of the human body are placed in danger (such as placing dry ice in the mouth or dipping hands into liquid nitrogen).

10. Do not use "open" containers of volatile, toxic substances (e.g., benzene, CCl_4, CS_2, formaldehyde) without adequate ventilation as provided by fume hoods.

11. Provide written procedure, hazard, and disposal information for each demonstration whenever the audience is encouraged to repeat the demonstration.

12. Arrange for appropriate waste containers for and subsequent disposal of materials harmful to the environment.

Background for Teachers

This section provides you, the teacher, with a brief overview of the pharmaceutical chemistry industry, Hoechst Marion Roussel's role in the industry, and a content review.

Overview of the Pharmaceutical Chemistry Industry

Some of the most important tools used in the medical profession today are chemicals which prevent and cure disease, control physiological processes and conditions, and alleviate discomfort. The development of these pharmaceutical agents, or drugs, is one factor that has helped to increase the life span of Americans by almost 30 years since 1900 and has allowed individuals with a variety of afflictions to lead longer, more fulfilling, enjoyable, and healthy lives.

The use of chemical agents to relieve pain and cure illness dates back to prehistoric times. However, the wider variety of pharmaceutical agents available to people in our modern society leads to a much greater use and dependence upon them in everyday life. In the pharmaceuticals industry, sales of prescription pharmaceutical agents have reached over $20 billion annually with an additional $9 billion spent each year on nonprescription agents. It is estimated that in the United States over 40% of the population uses an over-the-counter pharmaceutical product in any given two day period. Physicians, pharmacists, and even the everyday citizen must become knowledgeable about these chemicals to ensure that they are properly prescribed and used.

A pharmaceutical agent is broadly defined as any substance that has an effect upon the structure and function of the body. Since almost any chemical will have some kind of effect on an organism, we more commonly define a pharmaceutical agent as any chemical that is used to treat disease. Pharmaceutical agents are developed to control a variety of diseases. They are used to kill or control infectious organisms (e.g., quinine, penicillin, and streptomycin); to control, supplement, or substitute for various metabolic processes (insulin, adrenaline, anticancer agents, lipid-lowering agents, birth control pills and ointments); and to control the nervous system by stimulating (amphetamines), by depressing (alcohol), by tranquilizing (diazepam), by sedating (barbiturates), or by killing pain (aspirin, morphine, heroin, chloroform).

On average, 10–15 years of work and $250 million are required to discover, develop, and obtain approval for a prescription pharmaceutical agent. First, a chemist must become knowledgeable about a particular disease by completing a thorough library search. This includes searching the literature for general chemical structural characteristics of substances that might be suitable for inhibiting or activating the biomolecules of a particular metabolic process. Based on this information, an organic chemist attempts to synthesize one or more compounds with the desired characteristics. Once these compounds are purified, the chemist confirms the purity and structure of the compound by various analytical methodologies and instrumentation.

The effectiveness of these molecules in the treatment of disease must subsequently be determined. This involves determining the proper dosages and investigating the possibility of toxic side effects, a process that may take three to four years. If the initial investigations are positive, an application for a patent is submitted to the government. Patent rights last for 17 years from the date of issue.

Further biological testing of the molecule continues with a biochemist conducting experiments to confirm that the newly synthesized pharmaceutical agent will indeed affect the desired metabolic processes. Moreover, the biochemist must also determine the metabolic breakdown products and test them for possible toxic effects. With positive data on all the aforementioned tests, a scale-up of the synthesis is performed and enough of the compound made for clinical studies. Permission for clinical studies must be requested from the Food and Drug Administration (FDA) and a protocol developed so that a large enough pool of patients are examined to provide a valid statistical analysis of the product's effectiveness.

If all of these studies show that the pharmaceutical agent is safe and effective, and if economic studies indicate that the product can be marketed at a reasonable price so that a profit can be made, approval is sought from the FDA to market the pharmaceutical. The FDA is responsible for examining all of the data that the company has obtained since the initial synthesis of the compound by the synthetic organic chemist and ruling on the safety and effectiveness of the pharmaceutical. If the FDA rules favorably, the product is then given a brand name and marketed through contacts with medical professionals. Typically, the number of new pharmaceutical agents released in a year is well under 50.

Once a pharmaceutical product reaches the market, usually less than five years of the patent remain for the company to recover the approximately $250,000,000 cost of research and development. Once the patent runs out on the brand name product, other companies can start producing and selling the pharmaceutical agent as a generic product.

Individual pharmaceutical agents normally have several names under which they can be marketed. For example, Hoechst Marion Roussel Research Institute developed a widely used non-sedating antihistamine to which they gave the brand name Seldane. (See Figure 1.) The chemical name for this compound is α -[4-(1,1-dimethylethyl)phenyl]-4-(hydroxydiphenylmethyl)-1-piperidinebutanol. Its generic name is terfenadine. The generic name is agreed upon jointly by the FDA and the company that initially discovers, develops, and holds the patent on the prescription product. When the patent for a brand-name product expires, other companies are permitted to market formulations of the active ingredient under the generic name. These other companies can often sell the competing product at a very low price because they did not incur the expenses of developing and testing the product.

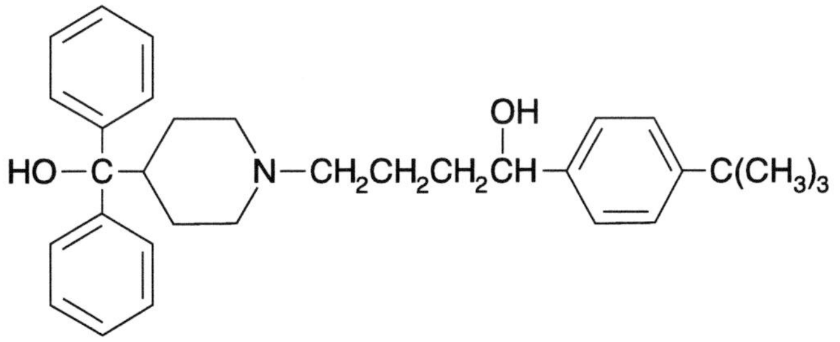

Figure 1: The structure of Seldane (terfenadine)

While both brand-name and generic products contain the same active ingredients, they may differ in the manner in which the product is packaged or formulated. Pharmaceutical formulations consist of the active ingredient and any other materials that may be added to produce the product or to aid in the delivery of the product to the proper biological site. The final product may be produced in a variety of delivery forms including, but not limited to, injectables, tablets, capsules, liquids, inhalants, and skin absorption pads. A generic pharmaceutical company is required to demonstrate that final delivery of the active ingredient to the proper biological site is equivalent to that of the brand-name product.

Some examples of materials that may be added in the final formulation are as follows:

- fatty acids or oils to facilitate the release of the product from the tableting or encapsulation machinery;

- various carbohydrates, talc, and oil mixtures to increase the bulk of the product or to hold together the tablet formulations;

- materials to coat the product to facilitate a timed release of the active ingredients;

- various coatings, dyes, or flavors to increase the public marketability of the product; and

- materials to carry the active ingredient to the proper biological site (for example, gases or solvents in an inhalant).

While each pharmaceutical product, generic or brand-name, may vary in its exact formulation, the active ingredients of both are identical, and both must show that the final formulation is beneficial and not harmful to the consumer. The generic company must also demonstrate that their product is equivalent to the approved brand-name product in the final delivery of the active ingredient to the proper biological site.

The Hoechst Marion Roussel, Inc. Research Institute (formerly Marion Merrell Dow Research Institute)

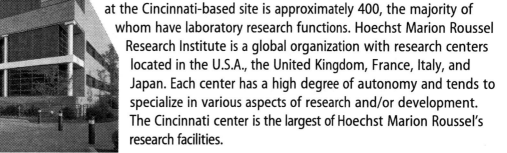

The Cincinnati Hoechst Marion Roussel, Inc. site, derived from the original William S. Merrell Company founded in 1828, consists of several research and administration facilities occupying over 400,000 square feet. The total number of research staff at the Cincinnati-based site is approximately 400, the majority of whom have laboratory research functions. Hoechst Marion Roussel Research Institute is a global organization with research centers located in the U.S.A., the United Kingdom, France, Italy, and Japan. Each center has a high degree of autonomy and tends to specialize in various aspects of research and/or development. The Cincinnati center is the largest of Hoechst Marion Roussel's research facilities.

The Hoechst Marion Roussel Research Institute in Cincinnati

Major products currently sold by Hoechst Marion Roussel, Inc. include

- **Seldane** (terfenadine), a family of widely used non-sedating antihistamines;
- **Rifadin** (rifampin), a family of antibiotics for tuberculosis treatment;
- **Lorelco** (probucol), an agent extensively used to lower cholesterol;
- **Norpramin** (desipramine), a standard treatment for depression;

A number of other products are available for overseas markets and are in various stages of clinical testing for approval in the United States.

The scientists at Hoechst Marion Roussel, Inc. are encouraged to be creative and to pursue new approaches to pharmaceutical agent design and testing. The unique structure of research at the company's Research Institute, which encourages biologists and biochemists to work closely with synthetic organic chemists and analytical chemists, creates a natural interplay that stimulates new ideas and research in the chemical and life sciences. Major findings from these cooperative efforts are used in the design and development of therapeutic agents. Major targets of research are illnesses affecting the respiratory, immune, and central nervous systems. Specific diseases targeted include atherosclerosis, asthma, diabetes, epilepsy, psychosis, arthritis, cancer, and AIDS.

A scientist reads a gel at the Hoechst Marion Roussel Research Institute.

This section provides a basic overview of some of the more important and complex content areas to be addressed in the activities. The overview assumes some familiarity with biochemistry; if you do not have this background, it may be necessary to refer to the biochemistry chapters of a general chemistry textbook for a more detailed explanation of these topics and structural notations. We begin by discussing polymers as biological molecules—proteins and enzymes, nucleic acids, and cellulose. Then we discuss chromatography as a method of analyzing components of complex mixtures. Finally, we examine the chemistry behind aspirin.

Polymers as Biological Molecules

Polymers are large molecules made of many repeating units called monomers. Polymers can be divided into two broad classifications: synthetic and natural. Synthetic polymers such as polyethylene, nylon, and polyester are used in many common products. Natural polymers such as proteins, nucleic acids, and carbohydrates are found in living systems.

Proteins, some of the most complex and numerous compounds found in living systems, are discussed later in this section. The two main types of nucleic acids are DNA (deoxyribonucleic acid) and RNA (ribonucleic acid). These large polymers are made of long chains of nucleotides and are discussed later in this section. Carbohydrates are composed of long chains of simple sugar molecules forming complex sugars and starches. These molecules are called polysaccharides and are used to store energy or to produce structural frameworks for cells. When a polysaccharide is broken down, the resulting simple sugars are used by the cells of the organism for energy. In plants, the carbohydrate used for structural material is called cellulose. Cellulose is the main constituent of plant products such as wood or cotton. Cellulose is discussed later in this section.

Proteins and Enzymes

Proteins are very large molecules and can be composed of hundreds of amino acids in a single chain or a combination of two or more chains held together by hydrogen bonds, ionic bonds, or disulfide bonds.

The properties of individual protein molecules depend on their shape and composition. The monomer unit of a protein is called an amino acid. There are 20 different amino acids found in living systems. Each amino acid consists of a basic, amine group ($-NH_2$) and a carboxylic acid group ($-COOH$) as shown in Figure 2. The R– represents a hydrogen or an organic side chain which varies from one amino acid to the next and gives each its individual characteristics. Amino acid monomers can react to form peptides of various lengths. The bonds joining the amino acid residues are called peptide bonds and are formed by the removal of water from two amino acid molecules during synthesis. (See Figure 2.)

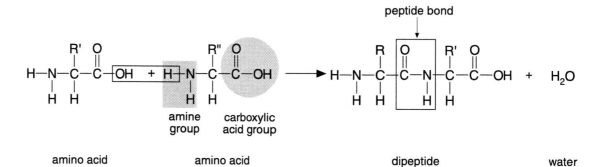

amino acid   amino acid   dipeptide   water

Figure 2: The general equation for the formation of a dipeptide

Long-chain polypeptides are often called proteins. These proteins have a specific shape and structure, which depends on the sequence of amino acid units from which they are made (including identity, relative numbers of individual amino acids, and arrangement). This is referred to as the protein's primary structure. This long chain of amino acids can in turn be twisted or folded to form the secondary structure. These twists and folds are held in place by intramolecular hydrogen bonds. Helices or pleated sheets are the two common forms of secondary structure. These secondary structures can, in turn, be folded over onto themselves to form the tertiary structure. The folds of the tertiary structure can be held in place by disulfide bonds (–S–S–), hydrogen bonding, salt bridges (ionic attractions between positively and negatively charged R– groups), and hydrophobic interactions. Three levels of protein structure are illustrated in Figure 3.

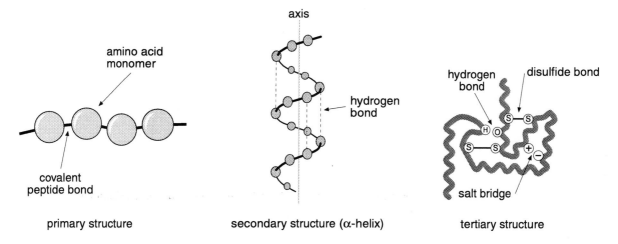

Figure 3: Three levels of protein structure

Proteins have molecular weights ranging from five thousand to millions of grams per mole. The function of a protein is determined by the sequence of its amino acids and its three-dimensional structure. In living systems, proteins take on structural, enzymatic, and regulatory roles. For example, they are necessary for clotting blood, transporting oxygen, regulating the speed of metabolic reactions, and determining the structure of the cells that make up our bodies.

Most proteins have structures that specifically recognize and bind a smaller molecule. If the structure of a protein is even slightly changed, the protein may no longer be able to bind this smaller molecule, which renders the protein inactive. This process is called denaturation. A protein can be denatured if it is exposed to certain outside influences such as heat, change in pH, or addition of certain chemicals. Figure 4 provides a generalized illustration of what occurs when a protein is denatured.

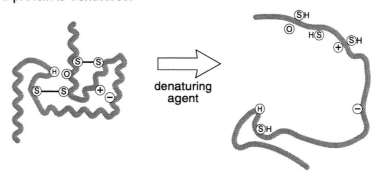

Figure 4: The effect of denaturation on protein structure

When proteins are denatured they lose not only their shape, they also lose their ability to function since the shape defines the function of the protein. An everyday example of protein denaturation can be seen when an egg is cooked. The contents of an egg are mostly protein. (Albumin is the major protein in egg whites.) When the egg is heated, the hydrogen bonds and disulfide bridges in the protein molecules are broken and the albumin changes from a clear liquid to a white solid.

Enzymes, sometimes referred to as biological catalysts, are a special group of proteins that regulate metabolism by controlling the speed of chemical reactions in the body. Enzymes are very specific; this means that a particular enzyme catalyzes the breakdown or synthesis of a particular molecule. One analogy of how enzymes work to catalyze reactions is called the lock-and-key model. The enzyme molecule fits into the substrate molecule (the specific molecule on which the enzyme will work) as a key fits into a lock. (See Figure 5.) Just as the proper key can open a lock, the correctly shaped enzyme can break down or put together specific molecules. Although enzymes are very large molecules, usually only a small region of the molecule (called the active site) is involved in the binding of the substrate. The molecular structure of the enzyme and substrate at the active site must be complementary for the enzyme to work.

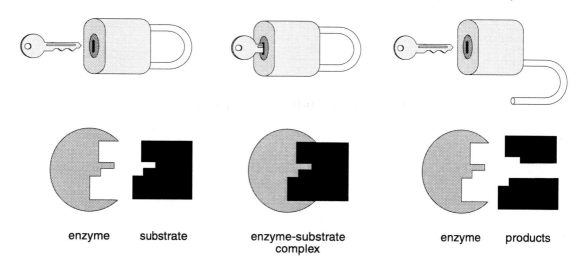

enzyme   substrate   enzyme-substrate   enzyme   products
                                complex

Figure 5: The lock-and-key analogy for enzyme activity

While the lock-and-key model of enzyme/substrate interactions is easy to visualize, it is considered to be greatly oversimplified for a variety of reasons. Biochemists actually view these interactions in terms of an induced-fit model. In this model, the binding of the substrate to the enzyme "induces" a change in the shape of one or both of these molecules, resulting in a shape that is different from the species that exist "free" in solution. (See Figure 6.) While neither the lock-and-key nor the induced-fit model explains the nature of the enzyme-substrate binding, the induced-fit model does help to explain enzyme activity better than the lock-and-key model does.

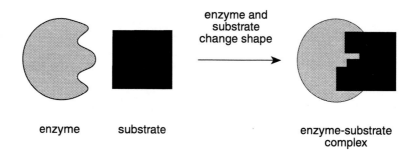

enzyme   substrate   enzyme and
                            substrate
                            change shape

                                            enzyme-substrate
                                            complex

Figure 6: The induced-fit model for enzyme activity

As previously stated, an enzyme recognizes a specific substrate or group of substrates with similar shapes. If the structure of the enzyme is even slightly changed, it usually will no longer recognize the substrate and therefore no longer function as a catalyst. It is also possible to have a molecule that resembles the substrate bind to the enzyme and thus prevent the enzyme from reacting with the substrate as expected. When this happens, the enzyme is said to be inhibited. Many pharmaceuticals work by resembling a biomolecule, binding to an enzyme that acts upon this biomolecule, and thereby preventing the normal (or in some disease states, the abnormal) reaction from occurring.

Nucleic Acids (DNA and RNA)

Nucleic acids—deoxyribonucleic acid (DNA) and ribonucleic acid (RNA)—hold the information for the reproduction, growth, and maintenance of an organism. Nucleic acids are natural polymers composed of complicated monomers called nucleotides. (See Figure 7.)

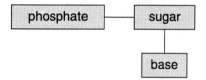

Figure 7: The general structure of a nucleotide

The structure of these monomers can be described as being composed of three parts. The central part of the monomeric structure is a sugar unit, either ribose or deoxyribose. (See Figure 8.)

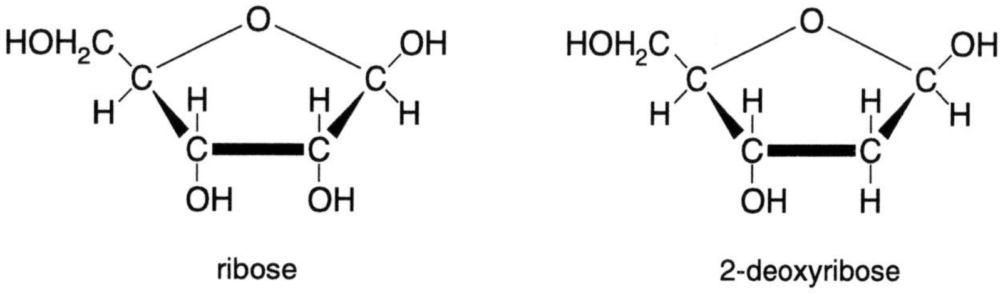

ribose   2-deoxyribose

Figure 8: Ribose and deoxyribose

The nitrogen-containing bases are shown in Figure 9. DNA and RNA both use cytosine, adenine, and guanine. However, in RNA thymine is replaced by uracil.

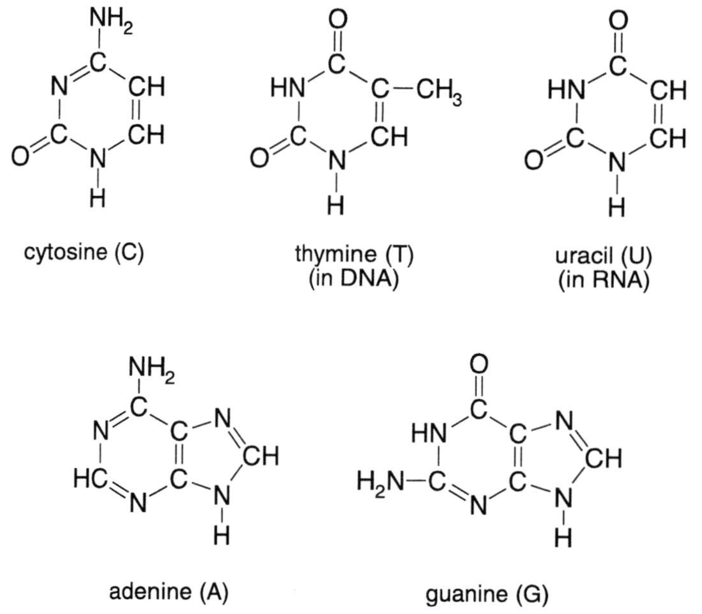

cytosine (C)
thymine (T) (in DNA)
uracil (U) (in RNA)
adenine (A)
guanine (G)

Figure 9: The nitrogen-containing bases in DNA and RNA

The sequence of these nucleotides in the DNA molecule determines the genetic makeup of the organism. The three-dimensional structure of DNA, the double helix, is composed of two polymerized strands of nucleotides wound around each other and held together by hydrogen bonds. Two views of the structure of DNA are shown in Figure 10. The darker lines in Figure 10a represent the bonded phosphate and sugar groups (the phosphate backbone) while the lighter bars represent the nitrogen-bases and the hydrogen bonding between the bases. Figure 10b shows these features in more detail. One of the four nucleotides in Figure 10b is shaded for clarity.

thymine
adenine
sugar
hydrogen bonds
cytosine
guanine

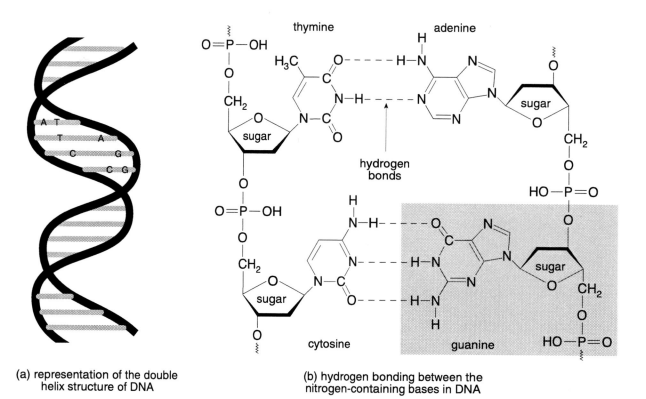

(a) representation of the double helix structure of DNA

(b) hydrogen bonding between the nitrogen-containing bases in DNA

Figure 10: The structure of DNA

Molecules of RNA have more varied biological functions than DNA. The nucleotides of RNA contain ribose instead of deoxyribose and the base uracil instead of thymine. Unlike DNA, RNA is not always double stranded; it usually consists of a single strand of nucleic acid. RNA is used to replicate or copy the DNA sequence so that all of the various proteins needed for metabolism and repair can be synthesized.

Cellulose

Cellulose is a condensation polymer of beta-D-glucose (β-D-glucose) units which combine by a dehydration process. This loss of water molecules, shown in the boxes in Figure 11, occurs between the hydroxyl groups (–OH) in adjacent β-D-glucose molecules. Alternate β-D-glucose molecules are inverted.

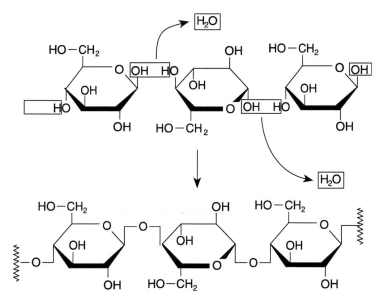

Figure 11: Condensation polymerization of β-D-glucose

There are about 10,000 glucose monomers forming each unbranched chain of cellulose. In the cellulose fiber, the cellulose chains exhibit both crystalline and amorphous regions. The cellulose chains are parallel to each other in the crystalline regions and more randomly spaced in the amorphous regions. Cross-linking is quite extensive because of the hydrogen bonding between the β-D-glucose monomers on adjacent chains. (See Figure 12.) This three-dimensional attraction provides much mechanical strength to the network and accounts for the structural role that cellulose plays in plants, including cell walls and stems.

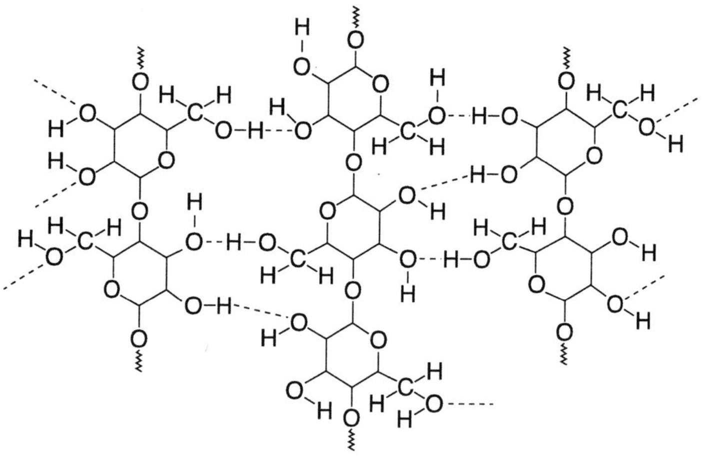

Figure 12: This simplified two-dimensional drawing shows some of the hydrogen bonding (designated by dashed lines) that can exist between cellulose chains.

Chromatography

Chromatography is the most widely used analytical method for separating the components of complex mixtures. Various types of chromatography exist, but they are all based upon the same simple principle: separation due to the differences in the attraction of a substance for two different media. All chromatographic procedures have two immiscible phases, a fixed or stationary phase over which the substances move and a mobile phase which flows over the stationary phase. The sample mixture is introduced into the mobile phase where it undergoes a series of interactions between the two phases. The differences in the interactions of each component in the mixture are governed by the physical or chemical properties of a particular substance. These interactions allow the substances to be attracted unequally to the stationary medium and thus flow across it at different rates. Strong interactions between a component and the stationary phase cause the substance to migrate more slowly through the mobile phase, while those components with weak interactions with the stationary phase tend to migrate more quickly. In this manner, substances are separated from one another.

Paper chromatography uses paper as the stationary phase and a solvent as the mobile phase. Paper is made of cellulose. Molecules which are attracted more strongly to the cellulose than to the solvent move more slowly on the paper than molecules that are less attracted to the cellulose. Molecules that are more attracted to the solvent flowing past the paper will move more rapidly than molecules that are less attracted to the solvent system.

Another chromatographic technique, referred to by various names (gel-filtration, gel-permeation, molecular sieve, or molecular-exclusion chromatography), provides a method of separating substances based upon their molecular size. Gel permeation chromatography (GPC) is a liquid chromatographic method which separates molecules primarily according to differences in molecular dimensions. Since the sizes of polymer molecules such as proteins, carbohydrates, and nucleic acids differ considerably, this chromatographic method is very useful in the separation of a complex mixture of these substances. The sample is introduced into the column and allowed to flow past a bed of rigid porous particles (the stationary phase). Molecules that have dimensions that allow them to fit into the pores are caught in the pores and detained in the column for a longer time than the larger molecules that cannot fit into the pores. The net result is that the large molecules come out of the column first and are separated from the smaller molecules that come out of the column later.

Aspirin Products

There are a variety of pharmaceutical agents available for relief from pain. Some of these products are available over-the-counter. Pharmaceutical agents that are used to relieve pain are called analgesics. The most common analgesic in the world is aspirin—acetylsalicylic acid. (See Figure 13.) In the United States, there are over 40 different prescription products and over 80 different nonprescription products on the market that contain aspirin. Furthermore, an estimated 20 billion tablets of aspirin or agent combinations containing aspirin are used each year. This is an average of over 80 tablets per U.S. citizen. About 30 million pounds of aspirin are produced annually in the United States with most of it being made by only two companies, Dow Chemical and Monsanto. Almost all of the companies producing aspirin-containing products obtain their aspirin directly from one or both of these two companies. It is not surprising then that independent studies have shown that all brands of aspirin are essentially the same and that the only major difference is the cost per tablet. Each aspirin tablet usually contains 325 mg of the active ingredient acetylsalicylic acid and an inert binder. "Extra-strength" formulations usually have 500 mg of the active ingredient but do not contain any additional active components.

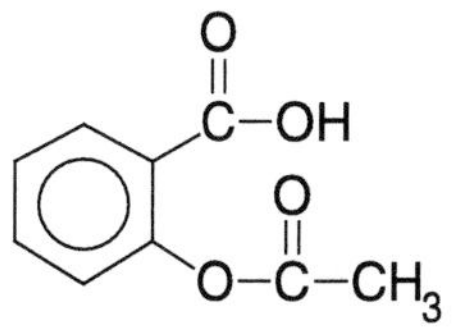

Figure 13: The structure of aspirin

Aspirin was first described in the science literature in the late 1700s when it was noted that mixtures of powdered willow bark relieved the pain symptoms caused by malaria. However, it wasn't until the late 1800s that the active compound, salicylic acid, was isolated from willow bark. Besides being a pain reliever, it was noted that salicylic acid also decreased fevers (an antipyretic) and reduced inflammation (an anti-inflammatory agent). The acidity of salicylic acid caused it to exhibit toxic effects in this purified form including ulcerations in the mouth and stomach.

Chemists attempted to modify the structure of salicylic acid to decrease its toxicity. The first attempt was to neutralize the acid by forming the sodium salt. Although the sodium salt did not irritate the mouth as much as the original salicylic acid did, it was still very irritating to the stomach. Further attempts to modify the structure eventually led to the esterification of the acid functional group. Several different esters were attempted before acetylsalicylic acid (more commonly called aspirin) was made. Acetylsalicylic acid is made by reacting salicylic acid with acetic anhydride. Not only was aspirin less acidic than the original salicylic acid, it also retained the analgesic, antipyretic, and anti-inflammatory properties of the original compound. Aspirin is also now known to inhibit the clotting of blood (an anticoagulant).

This method of modifying compounds to retain their desired effects and to remove their toxic properties is still the method used by modern-day pharmaceutical research groups. They are now aided by the increased technology of molecular modelling and by improved instrumentation to determine molecular structure.

Aspirin appears to relieve pain, reduce fever, and suppress inflammation by inhibiting the synthesis of prostaglandins. Prostaglandins are important hormone-like metabolic regulators that are also involved in regulating the response to pain. The interaction is with an enzyme that controls the rate of synthesis of prostaglandins. Aspirin binds to this enzyme and slows down the production of prostaglandins by changing the structure of the biomolecule.

Using the Activities in the Classroom

The activities in this module help students become aware of the importance of pharmaceutical chemistry and its relationship to the pharmaceutical industry.

Pedagogical Strategies

Before using an activity in the classroom, it is important that you read through it, try it, and adjust it according to the needs of your learners. Some of the activities (e.g., Activity 2, "Common Proteins: Reactions and Denaturations" and Activity 8, "Components of an Aspirin Tablet") lend themselves to inquiry-based or open-ended explorations. All of the experiments can be used in such a way that students cooperate in the learning process. Pair or group students into small teams for each activity.

Learning Cycle

You can use the activities "Cost/Mass Analysis of Different Aspirin Brands," "Components of an Aspirin Tablet," and "Making Aspirin and Oil of Wintergreen," to create a learning cycle. This learning cycle challenges students' preconceptions about aspirin as an over-the-counter (OTC) pharmaceutical while providing consumer and scientific information about this agent.

The first activity in the learning cycle, "Cost/Mass Analysis of Different Aspirin Brands," challenges students to test their notions about the effectiveness and value of name-brand aspirin samples. Students are asked to weigh samples of several types of aspirin, determine the actual mass of aspirin contained in each, and calculate the cost per tablet. Students learn that all regular-strength aspirins contain similar amounts of the same active ingredient, acetylsalicylic acid, as well as various other fillers. If told that only two companies in the U.S. make all the aspirin for almost every brand sold, students are surprised to learn that the cost per unit of aspirin in a name brand product may be twenty times higher than the cost for the same amount in a generic brand of aspirin! This activity also aids students in learning to read product labels and compare information given on the labels to the advertisements they hear or see promoting a product.

The second activity of the learning cycle, "Components of an Aspirin Tablet," affords students the opportunity to gather and analyze data about the composition and properties of various brands of aspirin. Students discover that aspirins contain binders, flavorings, buffers, and coatings as they analyze for pH, time required to dissolve in various solutions, presence of residual salicylic acid, and presence of binders.

The final activity of the cycle helps students expand their understanding of the chemistry involved in aspirin production and its relationship to oil of wintergreen. The activity "Making Aspirin and Oil of Wintergreen" allows students to prepare (or observe the preparation of) two compounds from the same starting material, salicylic acid, and subsequently analyze them for unreacted salicylic acid. Through reading and discussion, students discover that both chemicals have similar anti-inflammatory properties.

Science-Technology-Society (S-T-S) and Cooperative Learning

Challenge cooperative student groups to design a label and packaging for an over-the-counter product. The label must contain all needed information such as ingredients, weight, storage information, safety precautions, and directions for use. The package should be attractive yet practical. (Child-proof safety measures could be included here.) At the end of the project, each group can display their results. The class can then vote on the most effective and the most accurate designs and decide which product they would most likely be enticed to buy.

Cross-Curricular Strategies

- Mathematics classes can calculate the percent aspirin present in a pharmaceutical sample and the cost per unit of that aspirin. In chromatography experiments, students can compute R_f values and try to identify unknowns based on these values.

- Health classes can research the effect the mass production of antibiotics has had on world health and population growth. They may wish to start with penicillin, an antibiotic developed in the 1930s and '40s. The development of penicillin greatly changed the course of modern medicine.

- Students in health classes can research the use of OTC products in the U.S. today. This can lead into the importance of the pharmaceutical industry in the United States.

- Researchers are currently looking at natural sources for new pharmaceutical agents. For example, the cancer-treating agent, Taxol®, is found in the yew tree. Anti-clotting agents for the bloodstream have been found in the saliva of leeches. The growing awareness of the importance of maintaining biological diversity is influencing conservation efforts worldwide. Students may wish to research the development of pharmaceutical agents from such sources or look at the impact medical concerns have on conservation efforts.

- Social studies classes may wish to study the effects of pharmaceutical agents used to control diseases such as malaria on a given population. Population growth may occur, followed by food and resource shortages.

Individual or Group Projects

- Students can conduct a survey to determine the numbers and types of OTC products that are present in their homes. They can then poll their parents to find out what influenced them to buy a particular product brand (taste, cost, attractive package, effectiveness, advertisement on TV or in a magazine, etc.).

- Students can make posters from OTC product labels or from the advertisements for these products. The ads or labels can be analyzed for attractiveness, truthfulness, and readability.

- Students may want to compare ingredient labels from groups of OTC products other than analgesics (such as antacids). Antacids could also be tested for their ability to neutralize a given amount of a particular acid.

- Students can write research reports on the growth and function of the FDA and its efforts to maintain consumer safety.

- Students can research all the steps needed to bring a given pharmaceutical product to market from the original research, to testing, to its final marketing.

Annotated List of Activities and Demonstrations

To aid you in choosing activities for your classroom, we have included an annotated list of activities and demonstrations. This listing includes information about the grade level that can benefit most from an activity and a brief description of each activity. A Curriculum Placement Guide follows this list.

1. **Making a Polymer Foam** (middle to high school)
   Polyurethane foam is created from its two component parts. The emphasis of the activity can range from observations of the reaction to the actual chemistry involved in the synthesis of the polymer.

2. **Common Proteins: Reactions and Denaturation** (middle to high school)
   Common household items including egg albumin, sugar, salt, gelatin, corn starch, and meat tenderizer are tested for the presence of proteins.

3. **Speeding Up Elimination of Carbon Dioxide from the Blood** (middle to high school)
   Students use an enzyme found in chicken blood to speed up the decomposition of carbonic acid into carbon dioxide.

4. **Reactions with Potato Enzyme** (middle to high school)
   The enzyme tyrosinase is isolated from a potato. Using this enzyme, students perform experiments to determine some common properties of enzymes including reaction specificity and inhibition of function.

5. **Sizing Up DNA** (middle to high school)
   Students are introduced to the technique of gene splicing and gain insight into the relative sizes of a chromosome, a gene, and a bacterial cell.

6. **Paper Chromatography of Inks** (elementary to high school)
   Separation of the component pigments of the ink from black felt-tipped pens is achieved using inexpensive coffee filters and various solvents. The color separation that results provides a pattern which is characteristic of the ink tested.

7. **Separating Molecules by Gel Filtration** (middle to high school)
   Students use gel filtration to separate components of three different mixtures: a starch solution, tincture of iodine, and red creme soda.

8. **Components of an Aspirin Tablet** (middle to high school)
   Tests are performed on brands of aspirin to determine pH, time required to dissolve in various solutions, and the different components of the tablets such as starch (binder), residual salicylic acid, etc. Students discover that there are components in an aspirin tablet other than aspirin.

9. **Cost/Mass Analysis of Different Aspirin Brands** (middle to high school)
   Students weigh samples of several different brands of aspirin and calculate the percentage of aspirin in each tablet. They calculate the cost per tablet and the cost of 100 mg of aspirin for each brand. Recommended discussion topics include cost/mass analysis of extra-strength tablets versus regular-strength ones.

10. **Making Aspirin and Oil of Wintergreen** (middle to high school)
    Students prepare aspirin and oil of wintergreen from salicylic acid. The products are then used for further tests including analyzing the aspirin for unreacted salicylic acid.

Activities	Topics						
	Nature of Matter	Science and Technology	Scientific Method	Health	Mass, Volume, and Density	Chemical Reactivity	Biochemistry
1 Making a Polymer Foam	●	●	●		●	●	
2 Common Proteins: Reactions and Denaturation	●	●	●			●	●
3 Speeding Up Elimination of Carbon Dioxide from the Blood	●	●	●			●	●
4 Reactions with Potato Enzyme	●	●	●			●	●
5 Sizing Up DNA		●		●			●
6 Paper Chromotography of Inks	●		●				
7 Separating Molecules by Gel Filtration	●	●	●	●	●		●
8 Components of an Aspirin Tablet	●	●	●	●	●	●	●
9 Cost/Mass Analysis of Different Aspirin Brands			●		●	●	
10 Making Aspirin and Oil of Wintergreen	●	●	●	●	●	●	

Activities and Demonstrations

Making a Polymer Foam

How do you know when a chemical reaction has occurred? The reaction that takes place in this activity gives us several clues—the mixture expands to approximately 30 times its original volume, gives off heat, and becomes rigid. Students investigate several variables that affect this chemical reaction.

Recommended Grade Level .... 4–12 as a demonstration, 7–12 as a hands-on activity
Group Size .... 1–4 students
Time for Preparation .... 20 minutes
Time for Procedure .... 40–60 minutes (+ overnight for hardening)

Materials

Opening Strategy
Per Class
- 1 Tbsp vinegar
- 1 tsp baking soda

Procedure
Per Class
- 25–30 mL each of Part A and Part B of a polyurethane foam system
- 2 clear plastic cups
- wooden stirring stick or straw
- newspapers or paper towels
- (optional) food color
- (optional) paring knife
- plastic, disposable gloves
- utensil for measuring 10-mL and 25-mL increments to calibrate cups

Per Group
- 30 mL of Part A and Part B of a polyurethane foam system
- balance
- 4 clear plastic cups
- 2 wooden stirring sticks or straws
- 2 250-mL beakers or "pop beakers" made from 2 cut-off plastic 2-L soft-drink bottles
- thermometer
- goggles
- gloves

Variations
Per Class
- 3-oz cups
- large, latex balloon or rubber glove
- disposable aluminum bread pan
- empty ice cream cone
- sand

Resources

Parts A and B of a polyurethane foam system can be purchased from toy or craft stores such as Johnny's Toys, Cincinnati, OH, under the name Mountains in Minutes™ Polyfoam or from a chemical supply company such as Flinn Scientific, P.O. Box 219, Batavia, IL 60510-0219, 800/452-1261.

- polyurethane foam system—catalog # C0335 for Parts A and B

Safety and Disposal

Goggles and gloves must be worn when performing this activity. This demonstration or experiment should only be performed in a well-ventilated area. Persons with a history of respiratory problems or known sensitivity to organic isocyanates should avoid this activity. The liquid reagents used in this activity are irritating to the skin, eyes, and respiratory system. If Part A, Part B, or the mixture comes in contact with the skin, wash with soap and water. Avoid contact with the foam for about 24 hours, since there may be some unreacted starting material on the surface of the foam that can irritate the skin. After about a day of curing in a well-ventilated area, the hardened foam is safe to handle.

To store unused polyurethane reagents, cap the cans tightly. Use the solutions within one year of purchase; otherwise they may polymerize within the cans. If a non-food refrigerator is available, store the reagents there to extend their lifetime.

Getting Ready

Spread newspapers or paper towels over the table or lab bench where the activity is to be performed.

Calibrate two cups at 25 mL: Measure 25 mL water, pour it into each cup, and mark the water line. Calibrate four cups at 10 mL for each group in the same way. This will help in quickly obtaining an accurate measure of the amounts of Part A and Part B without using graduated cylinders.

Opening Strategy

Have students give examples of what has happened when they've mixed two liquids together. You might offer examples such as chocolate syrup and milk, oil and vinegar, or concentrated fruit juice and water. Ask the students if they noticed significant changes in volume, a temperature change, or any other unusual event. Challenge the students to make observations as you mix 1 tsp baking soda into 1 Tbsp vinegar. They should report that the foaming and bubbling that occurs is evidence of a chemical reaction. Challenge them to look for similar changes while doing the activity.

Procedure

Part 1: Teacher Demonstration

 Do this activity only in a well-ventilated area while wearing gloves and goggles. Read the Safety and Disposal section before proceeding.

1. Pour 25–30 mL of "Part A" into a clear, 25-mL calibrated cup. Have the students describe the properties of Part A and record them on the board. (Possible answers include gooey, smelly, viscous, thick, tan or pale yellow in color.)

2. (optional) Add 2–4 drops of food color and mix thoroughly.

3. Pour the same volume (25–30 mL) of "Part B" into a second clear, 25-mL calibrated cup. Have the students describe the properties of Part B. (Possible answers include gooey, smelly, viscous, thick, dark brown in color.)

4. Pour Part B into the cup containing Part A. Mix the two liquids together with a wooden stirring stick or straw until the color of the mixture is uniform.

5. Place the cup in the middle of the newspaper or paper towels and observe it for about 5 minutes.

 Because nothing seems to be happening initially, you may want to suggest that the experiment is not working, that the reaction must be very slow, or that maybe something is missing. Alternatively, you could discuss how chemicals A and B can react together to form the chemical C. You may also want to mention that this process is an exothermic reaction.

6. As the foam begins to "grow" out of the cup, you can use the stick to help shape it. However, the stick can become stuck to the foam if left in contact with it for too long.

 Do not handle the foam until it hardens because it usually contains some unreacted reagents that can irritate the skin. Do not let students handle the foam for 24 hours.

7. While the foam is drying, demonstrate its rigidity by tapping it with the stirring stick.

8. Once cured, the foam can be cut open with a paring knife to show the small bubbles inside. Point out the similarity to bread rising.

9. Ask the class to suggest possible uses for the foam. (Possible answers include insulation, packaging, crafts.)

Part 2: Student Activity

 Do this activity only in a well-ventilated area while wearing gloves and goggles. Read the Safety and Disposal section before proceeding.

Reaction at Room Temperature
1. Using 10-mL calibrated cups, repeat Steps 1–3 from Part 1, but use only 10 mL each of Part A and Part B.

2. Place paper on the pan of the balance to protect it. Set the two cups containing the liquids and the stirring stick on the pan of the balance and determine their combined mass. Record this mass.

3. Mix the two liquids together as described in Part 1, Step 4, and place the cups with the stirring stick in the mixture back on the pan of the balance. Observe the reaction.

4. After the foam has formed, determine and record the combined mass of the cups, stirring stick, and foam.

Reaction After Sitting Overnight

5. Repeat Part 2, Step 1 using two new 10-mL calibrated cups. Allow the liquids to stand open overnight.

6. The next day, mix the liquids together and allow them to react.

7. Compare the results with results obtained when parts A and B were not allowed to sit out overnight.

Variations

- Conserve materials in Part 2 by using 3-oz cups and 3 mL (½ tsp) each of Part A and Part B. This should produce enough foam to expand and fill the smaller cup.

- Pour the reactants into a balloon or rubber glove and tie off the end. Knead the balloon or glove to mix the reagents and observe.

- Mix the reagents in a disposable aluminum bread pan. As the foam ages and is exposed to sunlight, it will eventually become a bread-like golden brown.

- Fill an ice cream cone approximately ¼ full with sand. Add approximately 10–20 mL of each reagent and food color to the cone. As the foam expands, it will fill the top of the cone and begin to overflow the sides of the cone so that it looks like real ice cream.

Extension

- Polyurethane is often used as insulation. Challenge students to design an experiment to test whether or not their polyurethane is a good insulator.
  Carve a hole in the polyurethane that is 2–3 inches deep and large enough to hold the base of a cup. Fill a cup with warm water and place it in the hole. Set the cup in the polyurethane in a pan of ice and water. For a control, place an uninsulated cup of water in the same pan. Monitor the temperature in the cups for 5 minutes. (Alternatively, you could fill the cups with room-temperature water and place the assembly in a pan of hot water.) Graph the data as temperature on the y-axis and time on the x-axis. The plot for the polyurethane assembly should stay fairly flat if the polyurethane is a good insulator.

Discussion

- Ask students to classify the formation of the polyurethane foam as an exothermic or endothermic reaction.
  Heat is released, so it is an exothermic reaction.

- The mass of the polyurethane foam is less than the mass of the two starting liquids. Ask students to reconcile the difference in masses with the law of conservation of mass.
  When the foam forms, a gas is produced that makes it expand. Some of this gas escapes into the atmosphere. If the reaction was run in a closed system, the mass of the foam plus the gas would be the same as the combined mass of the two liquids.

- Compare the foaming action of the reaction at room temperature and after sitting overnight.

Explanation

The reaction that forms the foam is accompanied by several interesting changes that provide "clues" that a chemical reaction has taken place. The first clue is that a slight color change is observed as the two reagents are mixed and begin to react. A second clue is that the reaction is exothermic—it gives off heat. (While you should not touch the chemicals in this reaction, holding your hands close to the cup allows you to feel a slight increase in temperature. Students can also feel the bottom of the cup after the reaction has occurred.) A third clue is that the observed foaming is an indication that a gas may be produced. All of these observations are indications that a chemical reaction did occur.

The foam made in this reaction is polyurethane and belongs to a group of chemicals called polymers. Polymers are large macromolecules made of hundreds or thousands of repeating units called monomers. The reaction in this activity is classified as a condensation polymerization reaction.

The polymer-forming reaction (polymerization) between the two liquids can be represented visually with a kinesthetic demonstration. Begin by dividing students into two groups to simulate the molecules of the two liquids, Part A and Part B. The students in the Part A group can stand with both of their hands on their hips forming two "loops" where reactions can occur. (See Figure 1.) The students in the Part B group leave their arms hanging freely at their sides, leaving two active sites (free arms) where reactions can occur. In this simulation, the student "molecules" will combine to form a long chain that represents a small segment of the actual polymer chain produced in the reaction. The rules of the simulation dictate that the only way the student "molecules" are allowed to join together is for a free arm of a Part B molecule to hook with one of the loops of a Part A molecule. The reaction produces a long chain of alternating A and B units.

part A   part B   reacted A and B units

Figure 1: A human model representation of the reaction

The reaction continues until all of one of the parts is reacted or until the chain wraps around, allowing an end A to react with an end B on the opposite side of the molecule. The kinesthetic demonstration provides a basic overview of what occurs in the reaction; a more detailed explanation of the chemistry follows.

The main component of Part A is a polyol, an organic molecule with more than one alcohol functional group. The main component of Part B is a polyfunctional isocyanate. ("Polyfunctional" means the compound contains more than one functional group.) These two components react to form the rigid, crosslinked polyurethane. (See Figure 2. The R and the R' represent the repeating monomer units in the long polymer molecules.)

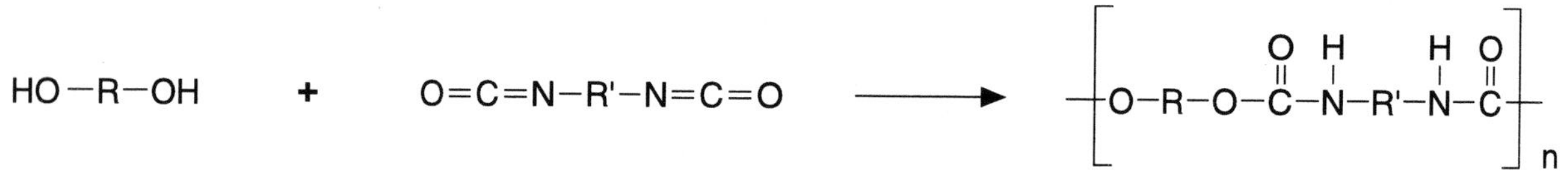

$$HO-R-OH \quad + \quad O=C=N-R'-N=C=O \quad \longrightarrow \quad \left[-O-R-O-\overset{\displaystyle O}{\underset{\displaystyle \|}{C}}-\overset{\displaystyle H}{\underset{\displaystyle |}{N}}-R'-\overset{\displaystyle H}{\underset{\displaystyle |}{N}}-\overset{\displaystyle O}{\underset{\displaystyle \|}{C}}- \right]_n$$

Figure 2: Formation of rigid polyurethane from a polyurethane foam system

Besides the polyol, Part A also contains a catalyst, a blowing agent (which was, as of late 1994, still reported to be a type of freon), and other minor components that are necessary for producing the foam. The heat generated by the reaction is enough to cause the blowing agent to vaporize, or enter the gas phase—this gas causes the foam to form. With time, the foam hardens as the long chains of polyurethane (formed during the reaction) crosslink with each other.

Key Science Concepts

- blowing agents
- evidence of a chemical reaction
- exothermic reaction
- phase change
- polymers and their properties

Cross-Curricular Integration

Language Arts
Have students write creative stories describing the reaction they observed.

Mathematics
Students can graph the results from the insulation extension.

References

Bailey, M.E., "Polyurethanes," *Journal of Chemical Education.* 1971, 48(12), 809–813.

Hocking, M.B.; Canham, G.W.R. "Polyurethane Foam Demonstrations: The Unappreciated Toxicity of Toluene-2,4-diisocyanate (TDI)," *Journal of Chemical Education.* 1974, 51(12), A580–581.

Pavia, D.L.; Lampman, G.M.; Kriz, G.S.; Engel, R.G. *Introduction to Organic Laboratory Techniques, A Microscale Approach;* Saunders College: 1990, pp 396–398.

"Polyurethane Foam;" *Fun with Chemistry: A Guidebook of K–12 Activities;* Sarquis, M., Sarquis, J., Eds.; Institute for Chemical Education: Madison, WI, 1993; Vol. 2, pp 89–93.

"The Preparation of Polyurethane Foam," *Chem Fax!* 1990, Publication Number 270.10; Flinn Scientific: Batavia, IL.

Shakhashiri, B.Z. *Chemical Demonstrations;* University of Wisconsin: Madison, WI, 1983; Vol. 1, pp 216–218.

Common Proteins: Reactions and Denaturation

Many important components of living organisms (enzymes, muscle tissue, hair, and nails) are made up of protein. In this activity, students examine the physical and chemical properties of some common proteins.

Recommended Grade Level 8–12
Group Size .... 1–4 students
Time for Preparation .... 20 minutes
Time for Procedure .... 60 minutes

Materials

Per Group
- 4 small test tubes with corks to fit
- piece of white office paper
- piece of clear plastic wrap
- newspaper
- about ⅛ tsp of each of the following:
  - NutraSweet™ (Equal®)
  - dry unflavored Knox® gelatin or lemon Jell-O®
  - table salt (sodium chloride, NaCl)
  - table sugar (sucrose)
  - corn starch
  - unseasoned meat tenderizer (e.g., Adolf's®)
- 2 tablets of glycine or another amino acid
- disposable plastic gloves
- 6 dropper bottles or droppers
- goggles

Per Class
- hot pot or coffee maker to heat water
- 1 egg
- cheesecloth or filter paper
- 50 mL 90–99% isopropyl alcohol
- tea bag
- 30 mL (2 Tbsp) unflavored Knox gelatin
- 50 mL 0.5% ninhydrin solution
- 50 mL tincture of iodine purchased or made as described in Getting Ready
- 4–5 beakers or containers sized to accommodate solutions in Getting Ready

Resources

The following can be purchased from a chemical supply company such as Flinn Scientific, P.O. Box 219, Batavia, IL 60510-0219, 800/452-1261.

- 0.5% ninhydrin solution—catalog # N0039 for 100 mL
- sodium iodate—catalog # S0218 for 25 g
- potassium iodide—catalog # P0067 for 100 g
- 3 M sulfuric acid solution—catalog # S0417 for 500 mL

The glycine or other amino acid tablets can be purchased at a health food store or pharmacy. The 90–99% isopropyl alcohol and tincture of iodine can be purchased at a pharmacy. Do not use the colorless form of tincture of iodine.

Safety and Disposal

Goggles should be worn when performing this activity. Ninhydrin solution is very flammable; keep it away from flames or heat which could cause it to ignite. Ninhydrin vapors are irritating to the eyes and respiratory system and the liquid is toxic if ingested. Use ninhydrin only in a well-ventilated area. Should contact with the skin or eyes occur, rinse the affected area with water for 15 minutes. If contact with the eyes is made, medical attention should be sought while rinsing is occurring. Ninhydrin reacts with the proteins in the skin and will stain them a dark blue-purple. Wear eye protection and gloves to protect the eyes and hands from the ninhydrin. Unused ninhydrin should be stored with other volatile solutions.

Isopropyl alcohol is flammable and intended for external use only. Avoid use around flames. Avoid contact with the eyes since damage can result. If contact does occur, rinse eyes with water for 15 minutes and seek medical attention. Goggles should be worn when using concentrated isopropyl alcohol solutions. Wash your hands after handling this chemical. No special disposal procedures are required.

None of the materials used in this activity are intended for consumption. Tea and iodine solutions can stain some materials and may temporarily stain the skin. Students should be warned to use caution while handling these materials. If you choose to make your own tincture of iodine, you must exercise caution when working with 3 M sulfuric acid.

Excess quantities of any of the solutions used in this experiment can be flushed down the drain with a large amounts of water. Unused tincture of iodine can be saved in a dark bottle for future use.

Getting Ready

To prepare 500 mL (2 cups) egg albumin solution: Gently stir one egg white in 500 mL water. Pour the solution through a cheesecloth or filter paper if a precipitate forms. Pour the clear solution into labeled dropper bottles.

To prepare 125 mL (½ cup) unflavored Knox gelatin solution: Dissolve 1 Tbsp Knox gelatin in 125 mL hot water and let cool to room temperature. Pour into labeled dropper bottles.

To prepare the tannic acid solution, steep one tea bag in 125 mL boiling water for 5–10 minutes. Remove tea bag and discard. Store the solution in a stoppered bottle.

Use caution when working with 3 M sulfuric acid. See Safety and Disposal.

If tincture of iodine is not purchased, 500 mL of solution can be made by dissolving 0.23 g sodium iodate ($NaIO_3$) in 125 mL water in a 1-L beaker. To the sodium iodate solution add 5 g potassium iodide (KI). Mix thoroughly until dissolved. To the sodium iodate/potassium iodide solution add 30 mL 3 M sulfuric acid (H_2SO_4). Add water to the resulting solution to bring the final volume to 500 mL. Place the tincture of iodine in a dark bottle for storage. For the activity, pour some of the tincture of iodine into labeled dropper bottles.

Opening Strategy

Discuss the roles proteins play in living things. For example, proteins provide structure (muscle), protection (skin, hair, nails), and allow us to digest food (enzymes). Use this information to challenge students to name things that contain protein (e.g., egg white, hair, gelatin). Explain that the chemical properties of a protein are determined in part by its molecular shape. This molecular shape is affected by temperature, pH, and exposure to certain chemicals. Tell students that in this activity they will test how various substances react with ninhydrin, a chemical used to detect the presence of amino acids, the building blocks of protein.

Procedure

Part 1: Testing for Proteins

1. Divide a blank piece of typing paper into eight sections with a pencil and label the sections with the words "NutraSweet," "gelatin," "egg albumin solution," "corn starch," "glycine" or "amino acid," "meat tenderizer," "salt," and "sugar."

2. Place a piece of newspaper on the desk. On top of the newspaper, place the labeled piece of office paper covered with a piece of clear plastic wrap.

3. Using the labeled sections of the paper as a guide, place two small piles (each about the size of an aspirin tablet) of each of the seven solids in their appropriate section. For the liquid egg albumin, use two drops of the egg albumin solution for each pile.

 Special caution must be used in the next two steps, which involve the use of ninhydrin and iodine. See Safety and Disposal.

4. To one of the piles of each labeled substance, add three drops of ninhydrin solution. (It takes approximately 30 minutes for the purple color to develop at room temperature.) After about 30 minutes, record any changes, especially specific color changes.

5. To the other pile of each labeled substance, add three drops of the tincture of iodine. Record any changes, especially specific color changes that occur immediately after the addition of iodine and after 5 minutes.

 Some amino acid tablets will give a positive test due to the presence of a starch binder in the tablet.

Part 2: Denaturing Proteins

1. Place about 1 mL (20 drops) egg albumin solution into a test tube. Place this test tube into a container of hot (near boiling) water for 5 minutes. Describe the changes that take place in the solution.

2. Place about 10 drops of the egg albumin solution into a test tube and add 30 drops of 90–99% isopropyl alcohol. Stopper the test tube and shake gently to mix. Describe the changes that occur, especially any physical changes and specific color changes.

3. Place 20 drops of egg albumin solution into a test tube and add 10 drops of tannic acid (tea) solution made in Getting Ready. Stopper the test tube and shake gently to mix. Describe the changes that take place in the solution, especially any physical changes and specific color changes.

4. Place 20 drops of Knox gelatin solution into a test tube and add 10 drops of tannic acid solution. Stopper the test tube and shake gently to mix. Describe the changes that take place in the solution, especially any physical changes and specific color changes.

Discussion

- Ask students why the ninhydrin test resulted in a faint purple color for proteins and a much darker purple color for the amino acid glycine.
  For equal weight samples of amino acids and proteins, amino acids will have many more amine groups available for reaction than proteins.

- Discuss what happens to proteins when they are denatured and cite specific ways proteins can be denatured.
  A protein becomes denatured when environmental conditions cause it to lose its shape. When a protein's shape is altered, it stops functioning. There are several factors that can cause a protein to denature: 1) high temperature; 2) change in pH; or 3) certain chemicals including alcohols or alkaloid reagents (such as tannic acid).

- Have students discuss how high temperatures and pH changes could be dangerous to proteins in the human body.
  A high fever can be dangerous because the enzymes and other proteins in the human body function over a narrow range of temperatures. Acids and bases can denature skin protein if they are spilled on the skin and can cause blindness if splashed into the eyes. Some forms of diabetes can cause the blood pH to change, causing a condition called acidosis.

Explanation

Proteins are important biochemical molecules. They act as the basic "building blocks" for many types of living tissue (e.g., muscle tissue, skin, and hair). Additionally, proteins are used to make the body's biological catalysts, called enzymes. The Content Review section of this module provides an overview of the basic structure of proteins and their role as enzymes.

A characteristic test for proteins and amino acids (the building block of proteins) involves the colorful reaction with ninhydrin. In Part 1, ninhydrin was reacted with NutraSweet, gelatin, egg albumin, glycine (or another amino acid), and meat tenderizer to possibly produce the characteristic dark purple color. (See Figure 1.)

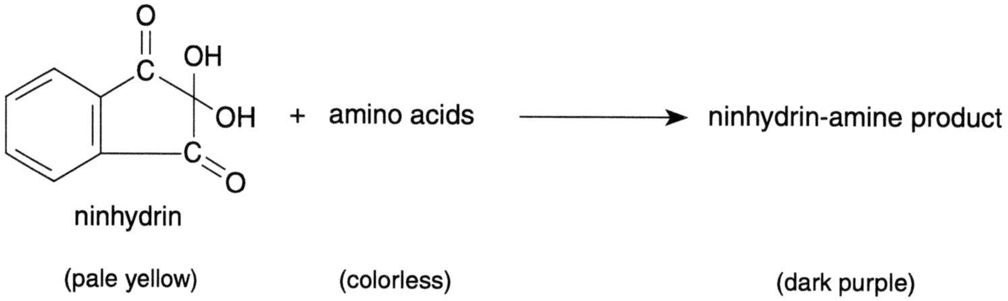

ninhydrin

(pale yellow)   (colorless)   (dark purple)

Figure 1: Reaction between ninhydrin and amino acids to produce dark purple complex

In contrast, the tincture of iodine (source of iodine) used in Part 1 did not react with protein. Instead iodine reacts with starch (in corn starch, meat tenderizer, and NutraSweet) to form a carbohydrate complex that is blue-black in color. (See Figure 2.)

$$\text{iodine} \quad + \quad \text{starch} \quad \longrightarrow \quad \text{starch-iodine complex}$$

(red-brown)	(colorless)	(blue-black)

Figure 2: Reaction between iodine and starch to produce blue-black complex

Corn starch is essentially pure carbohydrate, whereas meat tenderizer and NutraSweet contain both proteins and added carbohydrates. This is the reason they gave positive tests when treated with ninhydrin and with iodine.

As noted in the Content Review section, when proteins are denatured they lose not only their shape, they also lose their ability to function, since the shape defines the function of the protein. An everyday example of protein denaturation can be seen when an egg is cooked. The contents of an egg are mostly protein. (Albumin is the major protein in egg whites.) When the egg is heated, the hydrogen bonds and disulfide bridges in the protein molecules are broken and the albumin changes from a clear liquid to a white solid. This effect is demonstrated in Part 2 where the albumin solution is exposed to high temperatures in the hot water.

Proteins in the body can be denatured if the body temperature gets high enough. This is why high fevers are a cause for concern. The denaturing caused by the fever renders enzymes inactive and prevents them from carrying on important chemical reactions.

Acids, bases, alcohols, and alkaloid reagents can also denature proteins and can damage proteins in the human body if they are ingested or come in contact with the skin or eyes. It is important to wear goggles and gloves and to follow other safety precautions when working with these kinds of materials. The addition of concentrated solutions of an acid or a base changes the pH of the environment the protein is in. This can cause the protein to denature (due to the proton shift within the protein structure). Alcohol is used as an antiseptic and as a preservative for biological material because it will denature proteins in bacteria and kill them. Tannic acid is used to turn animal hides into leather because it converts the proteins in the skin into a form that is more resistant to decay. Tannic acid denatures the protein not because it is an acid (which would denature by changing protonation) but because the tannate anion in the solution forms ionic bonds with positively charged areas of the protein, disrupting salt linkages.

Key Science Concepts

- denaturation
- indicators
- protein structure
- reactions of carbohydrates
- reactions of proteins

Home, Safety, and Career
Make beef jerky and discuss how the chemicals change the protein structure to preserve it.

Life Science
Discuss proteins and carbohydrates, and their significance to the diet. Discuss protein structure and how the permanent wave process used by hair stylists (a "perm") changes the hair protein to produce the curls.

Social Studies
Have the students research how early buffalo hunters tanned buffalo hides and write a paper on the importance of buffalo to the Indians and the early settlers. (Tannic acid is used to tan hides by denaturing the skin protein and making it resistant to bacterial degradation.)

Speeding Up Elimination of Carbon Dioxide from the Blood

We inhale oxygen and exhale carbon dioxide (CO_2). But how? The answer involves a complex series of chemical reactions that utilize biological catalysts called enzymes. Without these enzymes, such reactions would be too slow to support life. This activity will show how a specific enzyme (carbonic anhydrase) can speed up the removal of carbon dioxide from the body.

Recommended Grade Level 7–12
Group Size .... demonstration
Time for Preparation .... 30 minutes
Time for Procedure .... 25 minutes

Materials

Opening Strategy
- 2–3 mL vinegar
- 3 10-oz clear plastic cups
- labels and markers

Procedure
Per Class
- 1-L bottle of seltzer water or club soda without citrate
- 20 g sodium hydroxide (lye, NaOH)
- 250-mL beaker
- bromothymol blue in solution
- 3 disposable, plastic pipets
- 100-mL graduated cylinder
- 2 10-oz clear plastic cups
- 2 stirring rods or spoons
- ice and ice bucket
- timing device with a second hand
- a few drops of blood from a package of fresh chicken livers
- gloves
- goggles

Variation
Per Class
- 1 of the following acid/base indicators in solution:
  - phenol red
  - red cabbage juice prepared in a blender from a head of red cabbage (See Getting Ready.)
  - bromocresol green

Extensions

Per Class

- 15 mL (1 Tbsp) vinegar
- ice

Resources

Bromothymol blue, phenol red, sodium hydroxide, and plastic, disposable pipets can be purchased from a chemical supply company, such as Flinn Scientific, P.O. Box 219, Batavia, IL 60510-0219, 800/452-1261.

- bromothymol blue solution—catalog # B0065 for 100 mL
- phenol red solution—catalog # P0100 for 100 mL
- sodium hydroxide solids—catalog # S0075 for 500 g
- Beral transfer pipets—catalog # AP8480 for 500

Phenol red indicator can also be purchased from a pool supply store or a pet store that sells aquarium supplies. Sodium hydroxide can also be purchased at a grocery or hardware store as household lye.

Safety and Disposal

Goggles and gloves should be worn when performing this activity. The solids and solutions of sodium hydroxide (NaOH) are very caustic. They can cause chemical burns and destroy cell membranes. Contact with the skin and eyes must be prevented. Should contact occur, rinse the affected area with water for 15 minutes. If the contact involves the eyes, medical attention should be sought while the rinsing is occurring.

Dissolving sodium hydroxide in water is an exothermic (heat-producing) process. The solutions can get very hot. Handle carefully. It is advisable to dissolve the sodium hydroxide near a sink so that water is readily available for rinsing in case any sodium hydroxide comes in contact with your skin.

All waste solutions from this activity can be diluted with water and flushed down the drain.

Getting Ready

Prepare a carbonic acid solution from a bottle of commercial seltzer water or club soda (without citrate). The solution should not be excessively carbonated, so release some of the carbon dioxide gas by opening the seltzer water or club soda and then reclosing it several times before using it for the experiment. After the bottle has been opened, it can be successfully used for at least 2–3 weeks if kept in a refrigerator. Place 150 mL of the solution on ice before the demonstration.

 A great deal of heat is evolved when dissolving sodium hydroxide in water. See Safety and Disposal for handling of sodium hydroxide.

Prepare 250 mL 2 M sodium hydroxide solution (NaOH) by dissolving 20 g solid sodium hydroxide in 200 mL water in a 400-mL beaker. After the solution has cooled to room temperature, add water until the volume reaches the 250-mL mark.

For the first extension, prepare 75 mL diluted vinegar solution (0.1 M acetic acid) by adding 15 mL (1 Tbsp) white vinegar to 60 mL (¼ cup) water. Place on ice before doing the demonstration.

For the second extension, prepare red cabbage juice by putting about two cups of coarsely chopped purple cabbage and ½–¾ cup of water in a blender. Blend on high speed until the cabbage is very finely chopped. Strain the mixture, reserving the deep purple liquid which makes an excellent acid-base indicator.

Opening Strategy

Label three clear plastic cups "water," "acid," and "base." Add 75 mL water to each cup. Add 2–3 mL (½ tsp) vinegar to the cup marked "acid," and add 2–3 mL 2M sodium hydroxide solution (2 M NaOH) to the cup marked "base." Add 10–15 drops of bromothymol blue to each cup and compare the colors. Emphasize the color of the indicator in acid (yellow), base (blue), and water (yellow/green).

Procedure

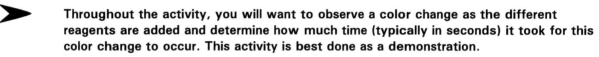

Throughout the activity, you will want to observe a color change as the different reagents are added and determine how much time (typically in seconds) it took for this color change to occur. This activity is best done as a demonstration.

Part 1: Neutralization of Carbonic Acid

1. Pour about 75 mL ice-cold carbonic acid solution into a clean plastic cup. Add 10–15 drops of bromothymol blue and stir to produce an easily visible yellow color.

2. Use a pipet to add about 2 mL (40 drops) 2 M sodium hydroxide solution (NaOH) without stirring. Note the color change.

3. Immediately begin timing to determine how long it takes for a second color change to occur.

The solution will not return to the same yellow color as above because of the carbonic acid/carbonate buffer system that is formed which keeps the pH between 6 and 7. At that pH, bromothymol blue indicator yields a pale yellowish-green color. The fading of the basic color is, however, still quite evident. The basic color usually persists for about 25–45 seconds.

Part 2: Using an Enzyme

1. Repeat Step 1 of Part 1 in another plastic cup with another stirring rod or spoon.

2. Add several drops of the chicken liver blood to the cold carbonic acid/bromothymol blue solution.

3. Use a pipet to add about 2 mL (40 drops) 2 M sodium hydroxide solution without stirring.

4. Immediately begin timing to determine how long it takes for a color change to occur.
The blue color usually persists for about 5–10 seconds. The final color will have a reddish-orange tint imparted by the blood.

Variation

- Perform the Procedure using phenol red (red in base), red cabbage juice (yellow in base), or bromocresol green (blue in base) instead of bromothymol blue. (The limitation of these other indicators is that they make it more difficult to detect color changes in the blood in Part 2.)

Extensions

- Repeat the Procedure replacing carbonic acid with dilute vinegar to provide a baseline comparison; the neutralization of most weak acids is very rapid.

- As a contrast, run this demonstration at room temperature. The main difference will be due to the fact that the solubility of the carbon dioxide will decrease as the temperature rises. With the release of the carbon dioxide, the enzyme may not work as fast due to a difference in pH from the reaction done in ice-cold seltzer water. It would be a good exercise for the students to try to explain why the enzyme works more slowly.

Discussion

- Carbon dioxide is produced by the cell as waste and passed into the blood. How does the blood get rid of the carbon dioxide? Does the reaction have to be fast or slow? Why? *The blood carries the carbon dioxide in the form of bicarbonate ion (HCO_3^-), which is the predominant carbon-containing ion at physiological pH. The bicarbonate ion is in equilibrium with carbonic acid (H_2CO_3). When the carbonic acid arrives at the lungs, the enzyme carbonic anhydrase converts it to carbon dioxide which is exhaled. This must be a very fast reaction to allow for all of the waste to be converted into carbon dioxide in the span of a few seconds that the blood spends in the lungs.*

Explanation

Metabolism is a series of chemical reactions that occur in our cells and involve the breakdown of complex molecules into simple ones to provide the food and energy needed to survive. Carbon dioxide (CO_2) is produced as a by-product of this process. The carbon dioxide diffuses from the cells into the blood where it dissolves via a hydration reaction to form carbonic acid. The carbonic acid in turn dissociates to form hydrogen ions and bicarbonate ions (HCO_3^-). The equilibrium that is established is shown in Figure 1.

$$CO_2\ (g)\ +\ H_2O\ (l)\ \underset{\text{dehydration}}{\overset{\text{hydration}}{\rightleftharpoons}}\ H_2CO_3\ (aq)\ \rightleftharpoons\ H^+\ (aq)\ +\ HCO_3^-\ (aq)$$

| carbon dioxide | water | | carbonic acid | | hydrogen ion | bicarbonate ion |

Figure 1: The hydration equilibrium of carbon dioxide

The carbon dioxide which is produced by metabolism in the cells is carried away in the blood primarily in the form of the bicarbonate ion. It is important that the body have a method to expel carbon dioxide and thus lower the H^+ and HCO_3^- levels in the blood to acceptable levels. If you were suddenly unable to exhale, the pH of your blood would decrease to dangerously low levels, a condition known as acidosis.

To reduce the H⁺ and HCO₃⁻ levels, the equilibrium must be shifted to the left to form CO_2 (g) via the dehydration process caused by removing the H_2O from the H_2CO_3 species. Dehydration is the reverse of the hydration reaction previously discussed. However, the dehydration process is very, very slow. If there were not a mechanism for speeding this reaction up, life as we know it would not exist. Fortunately, the blood contains the enzyme carbonic anhydrase, which catalyzes this reaction and makes it possible for the blood in the lungs to eliminate the carbon dioxide in about 1 second (the typical time of residence of the blood in the alveoli).

Enzymes are proteins that act as catalysts for the chemical reactions that take place in living things. A catalyst is a substance that affects the rate of a chemical reaction without being consumed in the reaction. Very small amounts of catalyst are needed as they are not used up or consumed by the reaction.

Carbonic anhydrase is a relatively small enzyme with a molecular weight of about 30,000. The catalyst can speed up either the hydration reaction of carbon dioxide or the dehydration of carbonic acid. Carbonic anhydrase is widespread in nature, being found in animals, plants, and certain bacteria. It is present in chicken blood as well as in human blood. Carbonic anhydrase has the highest activity of any enzyme known, with a turnover number of about 600,000 per second. This means that one molecule of enzyme converts 600,000 molecules of carbonic acid to carbon dioxide per second.

This activity was designed as a simulation of this enzymatic process. However, the blood was diluted by more than 500 fold from its initial concentration, and thus the increase in time to 5–10 seconds was observed.

In the activity, the carbon dioxide in aqueous solution (called the carbonic acid solution, H_2CO_3) is actually present primarily as dissolved CO_2 (aq). At a pH greater than 10, which is the approximate pH in the solution when the NaOH is first added in Part 1 (and 2), the neutralization of the OH⁻ with first the CO_2 (aq) and then the HCO₃⁻ accounts for most of the observations made. (See Figure 2.) The observed color change from blue to yellow/green eventually results as the pH falls towards the bicarbonate/carbonic acid pH of 6.4.

$$CO_2 (aq) + OH^- (aq) \longrightarrow HCO_3^- (aq)$$
carbon dioxide · hydroxide ion · bicarbonate ion

$$HCO_3^- (aq) + OH^- (aq) \longrightarrow CO_3^{2-} (aq) + H_2O (l)$$
bicarbonate ion · hydroxide ion · carbonate ion · water

Figure 2: The neutralization of a base (OH⁻) with dissolved aqueous carbon dioxide molecules and bicarbonate ions

The dilute vinegar used in the first extension shows that neutralization of most weak acids (e.g., acetic acid) is very rapid. The transient basic blue indicator color observed is due to inefficient mixing and not the rate of the reaction. For this reaction, there is no observable difference between this reaction carried out at room temperature and one using ice-cold acetic acid. However, the acid was used cold to show to the students that the low temperature is not the factor causing the slow change in the case of the carbonic acid solution. The carbonic acid solution had to be kept cold because the solubility of gases in solution decreases as temperature increases. Many times, biochemists will study reactions at lower temperatures in order to slow down the reactions for observation or to maintain proper concentrations of reactants.

Key Science Concepts

- acids and bases
- buffers
- catalysts and enzymes
- indicators
- rates of a reaction

Cross-Curricular Integration

Life Science

Discuss the chemistry of metabolism and respiration and explain the consequences of high levels of carbon dioxide in the body (acidosis).

Discuss the relationship between good health and enzyme activity. (For example, when a person has a high fever, how are enzymes affected?)

References

Bell, J. "Carbon Dioxide Equilibria and Reaction Rates: Carbonic Anhydrase-Catalyzed Hydration." Handout distributed by the author to the participants of the Woodrow Wilson National Fellowship Foundation Institute on High School Chemistry, July 1987.

Shakhashiri, B.Z. "Carbon Dioxide Equilibria and Reaction Rates: Carbonic Anhydrase-catalyzed Hydration;" *Chemical Demonstrations: A Handbook for Teachers of Chemistry;* University of Wisconsin: Madison, WI, 1985; Vol. 2, pp 122–126.

Reactions with Potato Enzyme 4

Enzymes are proteins that act as catalysts for chemical reactions in living organisms. Catalysts affect the speed of a reaction and are not consumed in the reaction. They can allow reactions to occur at a lower temperature—human body temperature, for example. In this activity, you will isolate an enzyme from a potato and study its function as a catalyst. This activity demonstrates how the function of a catalytic protein can be obstructed or inhibited by the presence of certain chemicals.

Recommended Grade Level 8–12
Group Size .... 1–4 students
Time for Preparation .... 15 minutes
Time for Procedure .... 60 minutes

Materials

Procedure, Part 1
Per Class
- medium-sized potato
- 1 Tbsp table salt (sodium chloride, NaCl)
- blender or homogenizer
- paring knife
- 60 cm (2 feet) of cheesecloth
- wide-mouthed container or 400-mL beaker
- plastic wrap or aluminum foil
- pan of crushed ice
- 1 dropper bottle, about 125–250 mL
- (optional) centrifuge

Procedure, Parts 2–4
Per Group
- 9 test tubes
- mug or medium-sized beaker of boiling or near-boiling water

Per Class
- 0.1% catechol solution
- 0.1% hydroquinone solution
- 0.1% resorcinol solution
- 0.2% sodium benzoate solution

 See Getting Ready for preparation of the four solutions listed above.

- 4 dropper bottles, about 125–250 mL
- goggles

Resources

The following chemicals can be purchased through a chemical supply company such as Sigma Chemical Company, P.O. Box 14508, St. Louis, MO 63178, 800/325-3010; or Flinn Scientific, P.O. Box 219, Batavia, IL 60510-0219, 800/452-1261.

- catechol—(Sigma C9510 for 100 g)
- hydroquinone—(Sigma H9003 for 100 g) (Flinn H0011 for 100 g)
- resorcinol—(Sigma R1000 for 100 g) (Flinn R0002 for 100 g)
- sodium benzoate—(Sigma B3375 for 100 g) (Flinn S0182 for 500 g)

Safety and Disposal

Goggles should be worn when preparing solutions and performing this activity. Handle catechol, resorcinol, hydroquinone, and sodium benzoate according to manufacturer's specifications. (See MSDS's.) These materials can be irritating to the skin, eyes, and respiratory tract. Avoid contact with the skin and eyes and use only in a well-ventilated area. Should contact occur, rinse the affected area with water for 15 minutes. If the contact involves the eyes, medical attention should be sought while rinsing is occurring.

All solutions produced in this experiment can be disposed of by flushing down the drain with large amounts of water. Any unused solutions can be saved for future use.

Getting Ready

Prepare an approximately 0.1% catechol solution by adding about 0.1 g catechol to 100 mL water. The measurements need not be exact. You may use a sample of catechol the size of half of an aspirin tablet in about ½ cup water. Pour the solution into a labeled dropper bottle.

Prepare an approximately 0.1% resorcinol solution by adding about 0.1 g resorcinol to 100 mL water. The measurements need not be exact. You may use a sample of resorcinol the size of half of an aspirin tablet in about ½ cup water. Pour the solution into a labeled dropper bottle.

Prepare an approximately 0.1% hydroquinone solution by adding about 0.1 g hydroquinone to 100 mL water. The measurements need not be exact. Use a sample of hydroquinone the size of half of an aspirin tablet in about ½ cup water. Pour the solution into a labeled dropper bottle.

Prepare an approximately 0.2% sodium benzoate solution by adding about 0.2 g sodium benzoate to 100 mL water. The measurements need not be exact. Use a sample of sodium benzoate the size of a small pea in about ½ cup water. Pour the solution into a labeled dropper bottle.

Procedure

Part 1: Isolation of the Enzyme Tyrosinase (Class Demonstration)

 This part of the Procedure should be carried out by the instructor and will provide enough enzyme for the entire class. For simplicity, this enzyme will be prepared in the absence of buffers and stabilizers. It is only usable for about 2 hours and should be stored on ice during this time.

1. Peel a medium-sized potato and cut it into golf-ball-sized pieces.

2. Place the potato pieces into a blender, and add 250 mL (1 cup) cold water and 1 Tbsp table salt (NaCl).

3. Blend for 20–30 seconds at medium speed.

4. Immediately filter the potato slurry through several layers of cheesecloth into a wide-mouthed container or beaker.

5. Cover the container with plastic wrap or aluminum foil and place the container in a pan of crushed ice. Allow the mixture to settle for 5 minutes.

6. Use the top (rather cloudy) portion of the solution as the enzyme extract for the subsequent parts of this experiment.

7. Centrifugation is not necessary to obtain reasonably good results. However, if a centrifuge is available, centrifuge at medium speed for about 30 seconds to separate precipitates (solids) from the mixture. Discard the solids.

8. Pour the enzyme extract into a labeled dropper bottle and store on ice until needed.

Part 2: Examination of Enzyme Specificity (Group Activity)

1. Label four clean test tubes A, B, C, and D and add the following liquids to the appropriate test tubes:

   Tube A: 10 drops water
   Tube B: 10 drops 0.1% catechol solution
   Tube C: 10 drops 0.1% resorcinol solution
   Tube D: 10 drops 0.1% hydroquinone solution

2. Add five drops of the enzyme extract to each of the tubes and *very gently* shake the tubes.

3. Examine the tubes and record observations after 5 minutes and again after 10 minutes. Use the codes in Table 1 to record the intensity of the color observed.

Table 1: Intensity of Color Change

Key	Color Intensity
–	no color change
+	faint color
+ +	definite color
+ + +	deep color

Part 3: Examination of Enzyme Inhibition (Group Activity)

1. Label three clean test tubes E, F, and G and add the following liquids to the appropriate test tubes:

   Tube E: 10 drops of 0.1% catechol solution and 10 drops of water
   Tube F: 10 drops of 0.1% catechol solution and 10 drops of 0.1% resorcinol solution
   Tube G: 10 drops of 0.1% catechol solution and 10 drops of 0.2% sodium benzoate solution

2. Add five drops of enzyme extract to each of the tubes and shake the tubes gently.

3. Examine the tubes and record observations after 5 minutes and again after 10 minutes. Use the codes in Table 1 to record the intensity of the color observed.

Part 4: Examination of Enzyme Denaturation (Group Activity)

1. Label two clean test tubes H and I and place 10 drops of enzyme extract to each.

2. Let tube H sit at room temperature for 5 minutes and place tube I into boiling (or near-boiling) water for 5 minutes.

3. Add 10 drops of 0.1% catechol to each of the tubes and shake the tubes gently.

4. Examine the tubes and record observations after 5 minutes and again after 10 minutes. Use the codes in Table 1 to record the intensity of the color observed.

Discussion

- Ask students what the color change in Part 2 indicated.
  The red color indicated that the enzyme was reacting with the catechol. Because a similar color change was not produced by resorcinol or hydroquinone, there is no evidence that a reaction occurred. Although the structures of the three molecules were similar, the enzyme was selective for catechol.

- Ask students to explain the differences in color of the solution mixtures in Part 3.
  Although there was catechol available to the enzyme in all three solutions, the resorcinol and sodium benzoate bound to the enzyme (without reacting with it), preventing the catechol from reacting with the enzyme.

- Ask students to use the results from Part 4 to predict what the effect of a high fever would have on the body.
  An elevated body temperature would denature some enzymes and cause them not to function. This is one reason why a high fever is dangerous.

- Discuss how enzymes work in the body.
  Enzymes work as biological catalysts in the body to allow complex chemical reactions to occur at lower temperatures and at a rate that supports life. A specific molecule (substrate) will fit into the active site of an enzyme. Only when the correct molecule is in the active site can the reaction occur. Sometimes similarly-shaped molecules can fit into the active site and cause the enzyme not to function. This process is called inhibition. Besides inhibition, changes in temperature and pH can also cause enzymes not to function correctly.

- Discuss conditions under which enzymes would stop functioning.
  Proteins such as enzymes are sensitive to conditions such as temperature and pH. An enzyme will also stop functioning if a molecule permanently binds to the enzyme's active site (inhibition).

Explanation

Enzymes are proteins that act as catalysts for the chemical reactions that take place in living things. Enzymes are very specific, meaning that a particular enzyme catalyzes the breakdown or synthesis of a particular molecule. (See the lock-and-key and induced-fit models for enzyme activity in the Content Review section.)

In Part 2 of this activity, three compounds with the same molecular formula, $C_6H_6O_2$, but slightly different structures are used. These compounds (called isomers) differ in the placement of one hydroxyl group (–OH). The structures of the molecules, catechol, resorcinol, and hydroquinone are shown in Figure 1.

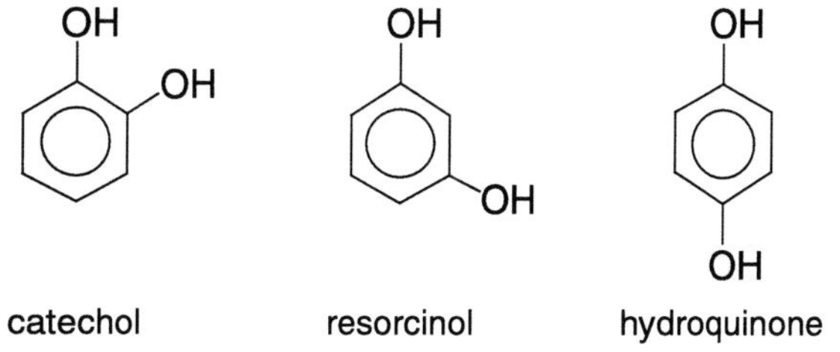

Figure 1: The structures of catechol, resorcinol, and hydroquinone

When catechol reacts with the enzyme tyrosinase (isolated from the potato), a multi-step reaction occurs, ending with the production of a polymerized red pigment that changes the color of the solution in the test tube. (See Figure 2.)

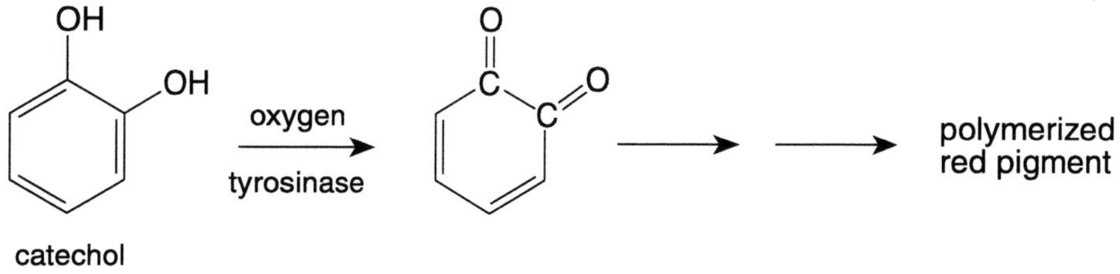

Figure 2: The reaction of catechol and the enzyme tyrosinase

Since these molecules are similar in shape, resorcinol or hydroquinone can sometimes "fool" the enzyme, by nearly fitting into the active site of the enzyme, tyrosinase. However, because the molecules do not have the exact shape required to fit into the active site of the enzyme, no reaction will occur. In other words, "look-alike" molecules have inhibited the enzyme reaction by filling the "lock" and not allowing the proper "key" to fit into the active site. A light green-brown color can appear with the resorcinol and the tyrosinase. This may be due to some impurity in the enzyme which can react with resorcinol. In Part 3, inhibition occurred in the mixture of catechol and resorcinol and in the mixture of catechol and sodium benzoate. Some brown color may develop in the catechol/sodium benzoate mixture due to the fact that some catechol may have reacted before the sodium benzoate inhibited the reaction. This activity showed that resorcinol, hydroquinone, and sodium benzoate, whose structure is shown in Figure 3, will compete with catechol for the enzyme.

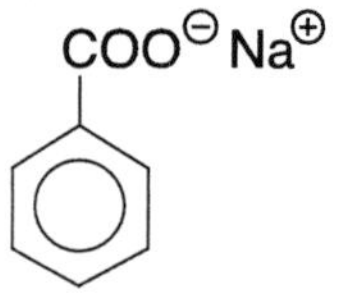

Figure 3: The structure of sodium benzoate

Enzymes, like other proteins, can be denatured, or twisted out of shape, by heat or certain chemicals. When this happens, the enzyme will no longer fit the substrate and the chemical reaction stops. This is one reason that high fevers are so dangerous—they denature some of the enzymes needed for vital biochemical reactions.

Key Science Concepts

- catalysts
- inhibition
- proteins and enzymes

Cross-Curricular Integration

Home, Safety, and Career
Have students discuss the effects of cooking food on proteins such as enzymes.

Life Science
Discuss the topic of enzymes with regard to biological systems. Explain that enzymes work as catalysts to make chemical reactions occur at lower (body) temperatures. This is how complex chemical reactions such as digestion can occur.

Reference

Baum, S.J.; Bowen, W.R.; Poulter, S.R.; Baumgarten, R.L. *Laboratory Exercises in Organic and Biological Chemistry,* 3rd ed.; MacMillan: New York, 1987; pp 241–248.

Sizing Up DNA

How is it possible to alter the basic material of life? Genetic engineering, once a topic of science fiction, is now a rapidly growing industry. Recombinant DNA technology offers many potential benefits to society. Students are introduced to the technique of gene splicing and gain insight into the relative sizes of a bacterial chromosome, a gene, and a bacterial plasmid.

> Recommended Grade Level 9–12
> Group Size .... Part 1, demonstration; Part 2, 4 students
> Time for Preparation .... 1 hour
> Time for Procedure .... 30–40 minutes

Materials

Procedure, Part 1

Per Class
- 3.8-m (about 12-foot) length of garden hose or aquarium tubing
- 30-cm length of wooden dowel rod that fits inside diameter of hose or tubing
- heavy-duty utility knife
- cutting board
- meterstick
- 2 index cards
- 4 different colors of tape

 Electrical tape (¾-in width) works well.

- transparent tape
- permanent marker
- (optional) sandpaper

Procedure, Part 2

Per Group
- 2 or more skeins of 3-ply knitting yarn

 The yarn should add up to a total length of at least 1,360 m.

- small package of yarn (a different color from the 2 skeins)
- meterstick

Getting Ready

1. Use the utility knife to cut the hose into eight pieces. The following quantities and lengths are needed: two 1.1-m pieces, four 30-cm pieces, and two 20-cm pieces.

2. Use the utility knife to score the dowel rod in six 5-cm segments. Use your hands to break the dowel rod apart at the score marks. If necessary, trim or sand off any splinters or sharp edges.

3. Assign colors of tape for each of the following bases: adenine (A), thymine (T), cytosine (C), and guanine (G).

Assemble the Bacterial Plasmid Model

4. Starting from the end and working inward, place bands of colored tape around each of the two longest pieces of hose in the pattern shown in Figure 1. (The bands of tape represent bases.)

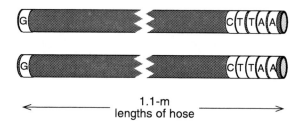

Figure 1: Place bands of colored tape in the pattern shown.

5. Place the hoses one on top of the other as shown in Figure 2. The G and C base pairs at each end should match according to the base pairing rules. (See Figure 3.) Tape the hoses together with transparent tape near the middle of the loop, not at the ends.

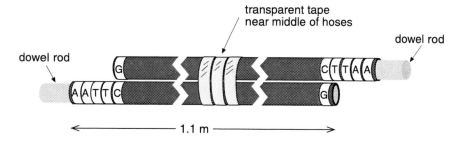

Figure 2: Orient the hoses as shown and tape them together.

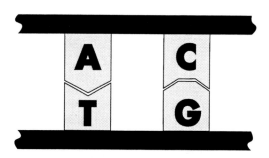

Figure 3: Base-pairing rules

6. Label each colored tape band with an A, T, C, or G according to the pattern in Figure 2. This pattern represents the recognition site of the restriction enzyme EcoRI.

7. Insert wooden dowel rod pieces in one end of each hose segment to enable the segments to be joined and separated. The separation between the G and the A represents the cleavage site of this enzyme. (See Figures 2 and 7.)

8. Label this setup as the bacterial plasmid.

9. Cover all the labels with transparent tape to prevent the letters from being rubbed off.

Assemble the Human Gene Model

10. Starting from the end and working inward, place bands of colored tape around two of the 30-cm hoses as shown in Figure 4.

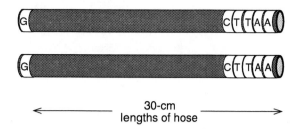

30-cm
lengths of hose

Figure 4: Place bands of colored tape in the pattern shown.

11. Orient the hoses as shown in Figure 5. Place these 30-cm pieces of hose one on top of the other so that the base pairing rules are followed. (See Figure 3.) Tape the hoses together with transparent tape as shown in Figure 5.

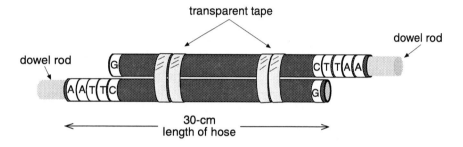

Figure 5: Orient the hoses as shown and tape them together.

12. Insert wooden dowel rod pieces in one end of each hose segment to enable the segments to be joined and separated.

13. Label this setup as the human gene.

14. Place bands of colored tape around the remaining 20-cm and 30-cm lengths of hose in the pattern shown in Figure 6.

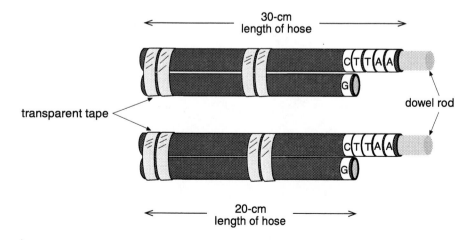

Figure 6: Place bands of colored tape in the pattern shown.

15. Place a 20-cm piece and a 30-cm piece together so that they line up at the ends without colored tape. The G and C base pairs should match according to the base pairing rules. (See Figure 3.) Tape them together near the middle and at the lined-up end. (See Figure 6.) Do the same with the other 20-cm and 30-cm piece.

16. Insert wooden dowel rod pieces in one end of each 30-cm hose segment to enable the segments to be joined and separated. (See Figure 6.)

17. Connect the two 20/30-cm DNA segments to the human gene as shown in Figure 7 to make a strand of human DNA.

18. Cover all the labels with transparent tape to prevent the letters from being rubbed off.

Represent the Enzymes

19. Label one index card as the restriction enzyme EcoRI (which cleaves DNA) and the other as the enzyme DNA ligase (which connects DNA).

Opening Strategy

Review the components of DNA. Also, review genes, chromosomes, and plasmids. Discuss with students how geneticists are able to isolate particular genes. Tell them that through this activity they will gain insight into genetic engineering (Part 1) and be able to conceptualize the actual size of genes (Part 2).

Procedure

Part 1: Recombinant DNA Demonstration

1. Place the hose representing the plasmid in a circle on a desk or table. Place the human gene in a line outside the plasmid. (See Figure 7.)

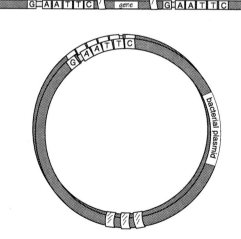

Figure 7: The plasmid and the human gene

2. Place the index card representing EcoRI in the center of the plasmid. Explain to students that EcoRI is a restriction enzyme that cuts DNA at a specific base pair recognition site. EcoRI cuts only at the site along the DNA at which pairs of the bases are arranged in the sequence GAATTC. EcoRI's staggered cut is shown in Figure 8.

3. Separate the human gene and plasmid as shown in Figure 8 to represent the cut made by EcoRI. Point out the GAATTC sequence of the bases around the cuts. EcoRI, the restriction enzyme used to cut the donor DNA, will also cleave the DNA of a host bacterium at a complementary site.

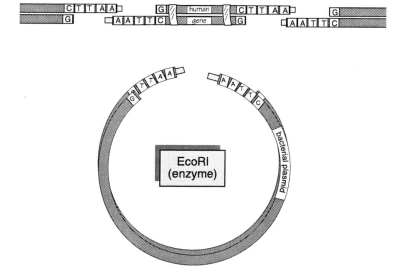

Figure 8: Separate the human gene and the plasmid.

4. Replace the index card representing EcoRI in the circle with the card representing the enzyme DNA ligase. Explain to students that DNA ligase is used to "glue" sections of DNA together.

5. Splice the piece of the human gene excised by EcoRI into the plasmid as shown in Figure 9. Explain to students that the plasmid now contains whatever genetic characteristics were present on the excised segment of the human gene.

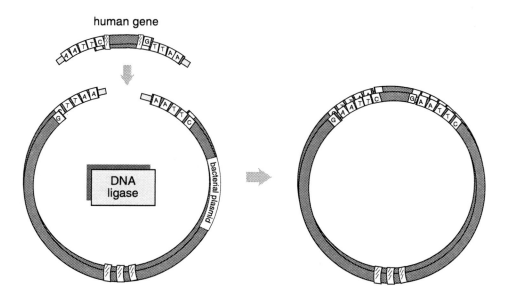

Figure 9: Splice the piece of human gene into the plasmid.

Part 2: Size Relationships of Genetic Material in a Bacterial Cell

1. Give each group of students the following information:

   - One base pair in DNA is 3.4×10^{-10} m long.

   - There are 1,200 base pairs in gene coding for an average protein containing 400 amino acids.

   - A plasmid for bacterial cloning has 3,000 base pairs and several genes.

   - Bacterial chromosomes have 4,000,000 base pairs and 3,000–4,000 genes.

2. Based on the information in Step 1, have students calculate the length of the following:

   - a single gene *(4.08 X 10^{-7} m)*;

   - a bacterial plasmid *(10.2 X 10^{-7} m)*; and

   - a bacterial chromosome *(1.36 X 10^{-3} m)*.

3. Explain to students that the class is going to assemble yarn models of the gene, the bacterial plasmid, and the bacterial chromosome. These models will help students understand the relative sizes of the three pieces.

4. Cut yarn from the small package into two pieces: one piece 40 cm long to represent a gene and another piece 100 cm long to represent the bacterial plasmid. Tie the 100-cm piece in a loop.

5. Unwrap the skeins of yarn and measure out a combined length of 1,360 m to represent the bacterial chromosome. (Do not tie the skeins together.)

6. Tie the 40-cm "gene" into the large mass of yarn (the "chromosome") by tying one end of the gene to the end of one skein and the other end of the gene to another skein. If you used more than two skeins, tie one end of each remaining skein to the end of the "chromosome" to form one long piece.

    You should now have one continuous length of yarn 1,360 m + 40 cm long. This length of yarn represents an entire bacterial chromosome.

7. Have two students hold the "chromosome," one at each end.

8. Have another student hold the "plasmid" beside the "chromosome."

Explanation

DNA, a large molecule that makes up genes, is the basis of all life. Genes are molecules that provide information and are used by organisms to carry out all of the functions necessary for the organism to develop appropriately and function properly. The DNA in each gene has codes (combinations of bases) that give the organism instructions to manufacture amino acids in the correct sequence for different proteins. A geneticist can create new sequences of bases that give different results than the original sequence would have produced. This is called genetic engineering.

Certain characteristics of DNA allow scientists to manipulate the stuff that genes are made of. The double helix structure is formed when two chains of DNA are linked together by base pairings. Base pairing is always the same, cytosine (C) always pairs with guanine (G), and adenine (A) always pairs with thymine (T). (See Figure 3 in the Procedure.) Because of these pairings, the order of bases of the opposite chain is predictable. If scientists know the pattern on one side of the double helix, the other side can be determined by pairing the bases according to the rules. Similarly, in the activity, you lined up the hoses by matching up the "bases."

Certain patterns of base pairs are recognized by proteins known as restriction enzymes. A restriction enzyme can break the DNA strand at exact sites between bases when the pattern matches the specific order that this enzyme recognizes.

In order for geneticists to accurately splice the DNA, the exact cutting site has to be known. Well-defined structures are important for predictable and repeatable results. There are hundreds of restriction enzymes with different site recognitions. Most recognize a pattern of 4 or 6 base pairs in a row. The one represented in this activity, EcoRI, recognizes the pattern GAATTC and always cleaves between the G and the A. (See Figure 8 in the Procedure.) The tubing representing the plasmid and human gene has exact breaks between these bases, and will always match up with the proper base pairs when put back together. Another enzyme is needed to put the ends of the chains back together. These enzymes are known as ligases, and they stick the DNA strands back together. The rules of base pairing are always followed. Scientists can use these predictable patterns to customize the recombined DNA.

In recent years, many new developments in genetic engineering have occurred. One of these developments is recombinant DNA technology. In recombinant DNA technology, DNA from one organism, the donor, is cut from the DNA sequence and placed into the DNA of another organism, the host. The organism that received the new DNA sequence will then become a carrier of this "new" sequence code and now has the ability to make a new protein according to this new DNA sequence, even if the organism could not have done this before. Recombinant DNA technology makes it possible for geneticists to take a section of human DNA and place it in bacterial DNA. The bacteria will then carry out the function called for by this section of human DNA. Bacterial plasmids are small circularly self-replicating bacterial DNA molecules. They are used as a transportation device to insert foreign DNA into host bacteria.

Key Science Concepts

- DNA
- genes
- genetic engineering
- recombinant DNA technology

Reference

Dillner, H. "Sizing Up DNA," *The Science Teacher.* March 1985, 53–55.

Paper Chromatography of Inks

Is black ink really black or is it a mixture of many colors? This activity demonstrates that black ink is really a mixture of pigments that can be separated using a technique called chromatography. Inexpensive items such as coffee filters, fingernail polish remover, and rubbing alcohol are used in this fun activity.

Recommended Grade Level 4–12
Group Size .... 1–4 students
Time for Preparation .... 10–20 minutes
Time for Procedure .... 20–30 minutes

Materials

Opening Strategy
- white paper towel
- clear plastic cup
- water-soluble markers

Procedure
Per Group
- 10–30 mL of 1 or more of the following solvents:
  - rubbing alcohol (70% isopropyl alcohol)
  - 0.1% table salt solution (sodium chloride, NaCl) (See Getting Ready.)
  - acetone-based fingernail polish remover

➤ **Be sure that each solvent is used by at least one group.**

- 10-oz clear plastic cup (Substitute a glass container if acetone-based remover is used as acetone may dissolve a polystyrene cup.)
- 1 or more of the following pieces of paper:
  - filter paper at least 15 cm in diameter
  - coffee filter
- 3–4 different brands of black, water-soluble, felt-tipped pens and markers

➤ **Crayola® and Vis-a-Vis® brands work well.**

- (optional) 13-cm x 13-cm (5-in x 5-in) piece of plastic wrap or plastic sandwich bag
- metric ruler calibrated in millimeters
- scissors
- pencil
- paper towels
- goggles

Variations
All of the materials listed for the Procedure, plus the following:
- M&M's®, Reese's Pieces®, and Skittles® candies
- food color
- Kool–Aid® powdered drink mix

- grass or leaves
- skins of fruits and vegetables
- permanent markers
- T-shirt material
- cheap, uncoated chalk
- paper shavings
- disposable pipets
- ink or dye

Resources

Filter paper can be purchased from a chemical supply company such as Flinn Scientific, P.O. Box 219, Batavia, IL 60510-0219, 800/452-1261.

- filter paper—catalog # AP3105, 15-cm diameter, for 100 sheets

Safety and Disposal

Goggles should be worn when working with acetone, a component of fingernail polish remover. Acetone is very flammable. Keep it away from flames and heat that could cause it to ignite. Acetone vapors are irritating to the eyes and respiratory system, and the liquid is toxic if ingested. Use acetone only in well-ventilated areas.

Getting Ready

To prepare a streaked paper towel for the Opening Strategy: Place several different color dots from water-soluble markers about 2 cm from the bottom edge of a white paper towel. Twist or fold the edge of the towel with the colored dots into a point and, making sure not to submerge the dots, dip the point of the towel into a cup containing a small amount of water (about 10–15 mL). Allow the water to travel up the towel past the dots until the colors begin to separate. Don't worry if the colors smear as long as there is a characteristic color pattern.

To prepare an approximately 0.1% table salt solution, dissolve 1.0 g (⅛ teaspoon) table salt (NaCl) in 1 L water.

Opening Strategy

Show students a paper towel streaked with color from a water-soluble marker as described in Getting Ready. Ask students how the towel might have been colored. Explain that most markers are made of a mixture of pigments. The procedure used in this activity allows the components of ink to be separated.

Procedure

 To provide variety and add to the class discussion, have each group use a different solvent to perform the activity and compare results as a class.

Part 1: Preparing the Filter Paper

1. Cut a piece of filter paper or a coffee filter into a 10-cm x 10-cm (4-in x 4-in) square.

2. Use a pencil to write the name of the solvent being used at the top of the filter paper.

3. Depending on the number of markers to be tested, divide the paper into three or four sections. To divide into three sections, fold the square into thirds. To divide the square into four sections, fold the square in half and then in half again to make accordion-type pleats. (See Figure 1.)

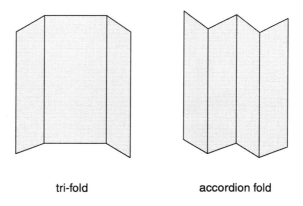

tri-fold   accordion fold

Figure 1: The folded filter paper

4. Use a pencil and ruler to draw a line across the filter paper perpendicular to the fold lines 1.5–2 cm from an edge. Treat this edge as the bottom of the paper.

5. Using a pencil, label the bottom of each pleated segment (below the line drawn in Step 4) with the brand name of the felt-tipped marker to be used on that segment.
➤ **DO NOT use marker for labeling the filter paper.**

6. Use the appropriate brand of marker to make a small dot of ink (smaller than the diameter of a pencil eraser) in the middle of the pencil line in the appropriate pleated segment of the paper.

7. Repeat Step 6 using a different marker on each segment of the paper.

8. Refold the spotted filter paper square along the original fold lines so that it will stand up with the ink spots near the bottom.

Part 2: Developing the Chromatogram

1. Add 10 mL (2 tsp) of the solvent to the plastic cup.

2. Stand the paper outside the cup. Be sure that the ink spots will be above the solution; if too much solution was added, pour some out before continuing.

3. Carefully stand the folded filter paper in the cup containing the solvent as shown in Figure 2. To develop the chromatogram more rapidly, cover the top of the cup with plastic wrap or invert a plastic sandwich bag over the cup.
➤ **Watch as the solution "creeps" up the paper; the components of the ink rise along with the solution, some moving faster than others. The development time depends on the kinds of paper and solvents used.**

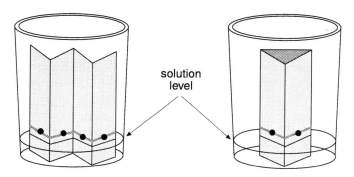

Figure 2: The paper standing in solution

4. Remove the paper from the cup when the location of the solution level (solvent front) reaches about 2 cm from the top edge of the filter paper.

5. Carefully lay the filter paper on a clean, dry paper towel. If R_f values (See Part 3) are to be determined, immediately mark the location of the solution level (solvent front) on the filter paper with a pencil.

Part 3: Analyzing the Results

The chromatograms produced in Part 2 can be analyzed to find out which colors of ink were used to make the different brands of ink.

a. For lower-grade students:

1. For each ink spot, list the component colors that separated starting at the top of the chromatogram and working downward.

2. Prepare a table of results for each brand of felt-tipped marker, listing the colors found in the chromatogram.

3. Compare the chromatograms of the different brands of markers to determine similarities. (For example, is blue always the fastest-traveling color? How do the pigments used in the different brands of markers differ in color composition? Are there one or two colors that are present in all the different brands of markers?)

b. For upper-grade students:

In addition to the analysis performed in Part 3a, students can make a rough calculation of the retention factor, R_f, for each pigment. For this activity, the retention factor may be defined as the distance the color component traveled divided by the distance that the solvent traveled.

1. Measure the distance from the middle of the original ink spot to the edge of the solvent front marked in Step 5 of Part 2. This is the distance that the solvent traveled.

2. Measure the distance from the middle of the original ink spot to the approximate middle of each streak of color. This gives an approximate value for the distance that each pigment traveled.

3. Calculate R_f for each pigment.

$$R_f = \frac{distance\ solute\ (color)\ moved}{distance\ solvent\ (acetone)\ moved}$$

4. Repeat Steps 1–3 for the other ink spots. Compare the R_f values for the same color found in the different brands of markers.

Variations

- For upper-grade students, it may be interesting to develop chromatograms using two different solvents for comparison. Have half of the groups use a moderately polar solvent, such as fingernail polish remover or acetone, and have the other half use an ionic solution, such as 0.1% salt water. The procedure for developing the chromatogram is the same for both solvent systems. The difference occurs in the order of separation of the pigments in the marker ink. Due to the differences in the polarity of the solvents, the order of the pigments in the salt solution may be different than the order of the pigments in the acetone. Have the students compare their chromatograms. Ask them to try to account for any differences between their chromatograms. This would be an excellent way to lead into or reiterate the concept of polarity.

- Use other types of pigments and colorful mixtures:
  - Candy coatings from M&M's, Reese's Pieces, and Skittles— Dissolve the coating from the candy with minimum amount of water (2–3 drops), dot the color coating on the paper, and use a 1% salt solution as the chromatographic solvent.
  - Food colors—Use a 1% salt solution as the chromatographic solvent.
  - Kool-Aid drinks—Make a concentrated slurry of the powder in water and use a rubbing alcohol/water mixture as the chromatographic solvent.
  - Grass or leaves—Extract the chlorophyll from the leaves using hot rubbing alcohol and use acetone as the chromatographic solvent.
  - Skins of fruits and vegetables—Extract the pigments by making a pureé in a blender. Spot the colored liquid and use a 0.1% salt solution as the chromatographic solvent.

- Perform radial chromatography. Place a spot of ink in the center of a piece of circular filter paper. Add solvent and the ink will separate outward from the center.
  - Water-soluble markers—Use water as the chromatographic solvent.
  - Permanent markers in all colors on T-shirt material—Use rubbing alcohol as the chromatographic solvent.

- Use media other than paper:
  - Cheap, uncoated chalk—Use chalk just like the paper. Place a spot of ink about 2 cm from the bottom of the stick. Place the chalk, with the ink spot down, in a clear plastic cup filled with about 10 mL of solvent. Cover. It should develop like the paper did, with the color separating as the solvent moves up the chalk.
  - Column chromatography (disposable pipets packed with chalk or paper shavings)— Place ink or dye at the top of the column and pour the solvent (acetone or acetone/water) through until you see bands of color appear. This technique allows for the separation and collection of the components of the dye mixture.

Extension

- Because the manufacturers of black markers use different combination of pigments, challenge the students to develop a method of identifying which black marker was used to produce a prepared chromatogram.

Discussion

- Ask students to explain how the pigments travel up the filter paper and what determines the rate at which the pigments travel.
  As the solvent travels up the paper, the pigment molecules travel with it. The different pigments in the ink move at different speeds based on their differing attractions for the solvent.

- Discuss the order of pigments in the different solvents.
  The most polar pigments travel the farthest in the 0.1% salt-water solvent. (Polar solvents are very good at dissolving polar molecules.) The separation sequence typically changes when acetone (a less polar solvent than water) is used.

Explanation

Although black ink may appear to be a single color, most black inks are comprised of a mixture of pigments, sometimes as many as eight or nine. (A pigment is a substance that imparts color to a material.) Paper chromatography can be used to separate the component pigments from one another. In the pharmaceutical chemical industry, paper chromatography is used to purify and identify substances throughout the development of new medications.

Paper chromatography separates mixtures based on the interaction of the samples with the solvent and the paper. The solvent travels up the paper (or other media) via capillary action or "wicking." As the solvent moves up the filter paper, it carries with it the pigments in the ink. Each pigment travels at its own speed depending on its interaction with the solvent (mobile phase) and the paper (stationary phase). Differences in these interactive forces among the various pigments cause some pigments to move up the paper at different rates. The pigments that are more soluble or are strongly attracted to the solvent move up the paper at a faster rate than those that are less soluble and have a smaller degree of attraction for the solvent. This separation of pigments becomes more apparent as the chromatogram is allowed more time to develop.

A closer look at the chromatogram indicates that there is more going on than just pigments travelling at different rates. Some of the pigments end up in very concentrated zones; others are spread out in diffused bands. Some pigments tend to streak out in lines; others tend to form fanlike patterns. These effects are very likely a result of the intermolecular forces that exist between the molecules of the various pigments.

The retention factor, R_f, for a given pigment is the ratio of the distance traveled by a given pigment divided by the distance traveled by the solvent such as acetone/water solution. For example, if a pigment has an R_f value of 0.5, then the pigment moved half as far as the solvent did. Factors such as porosity of the paper, thickness of the paper, and the type of solvent would affect the values for R_f. This value allows you to compare colors from one sample to the next to determine if there may be similar components in the different samples.

Key Science Concepts

- chromatography
- mixtures
- pigments
- polarity
- separation methods

Cross-Curricular Integration

Art
Create colorful bookmarks or coasters by laminating some of the students' chromatography samples.

Language Arts
Crime labs use chromatography to separate the components of a substance in order to identify it. For example, chromatography might be used to identify the ink used to write a ransom note. Using this information, write a story in which the police/private detective uses chromatography to save the day.

Mathematics
Older students practice making accurate metric measurements. They also use arithmetic to calculate the R_f values for each pigment found on the chromatogram.

References

"Colorful Separations;" *Fun With Chemistry: A Guidebook of K–12 Activities;* Sarquis, M., Sarquis, J., Eds.; Institute for Chemical Education: Madison, WI, 1992; Vol. 2, pp 1–35.

Mewaldt, W.; Rodolf, D.; Sady, M. "An Inexpensive and Quick Method for Demonstrating Column Chromatography of Plant Pigments in a Spinach Extract," *Journal of Chemical Education.* 1985, 62(5), 530–531.

Scharmann, L.C. "Autumn Leaf Chromatography," *Science and Children.* 1984, 22(1), 11–12.

Separating Molecules by Gel Filtration

Sometimes scientists need to separate one particular chemical from a mixture. How do they do this? Scientists have developed techniques that separate chemicals based on the size of molecules. One such technique is gel filtration. In this activity, students use gel filtration to separate components of a starch and tincture of iodine.

Recommended Grade Level 8–12
Group Size .... 1–4 students
Time for Preparation .... 30 minutes (+2 hours for first-time hydration)
Time for Procedure .... 50–60 minutes

Materials

Opening Strategy
- 1-ft x 3-ft piece of wire screen with ½-in square openings
- 1-ft x 3-ft piece of stiff cardboard
- 4 Styrofoam™ balls of each of the following diameters: 1½-in, 1-in, ¼-in

Procedure, Parts 1, 2 and 3
Per Group
- disposable glass Pasteur pipet
- disposable plastic pipet
- 3 mL hydrated Sephadex G-50 80–100 mesh resin (See Getting Ready.)
- small piece of tissue or filter paper
- 400-mL beaker
- cup or beaker of water
- empty cup or beaker
- 6 small test tubes
- about 40 drops of liquid laundry starch or 2 Tbsp corn starch
- about 60 mL tincture of iodine purchased or made as described in Getting Ready
- 2 plastic dropper bottles with lids
- (optional) a piece of wire 6–8 inches long (e.g., a large unfolded paper clip)
- (optional) test tube rack
- (optional) timing device with a second hand
- goggles

Procedure, Part 3
Per Group
- 10-oz disposable cup
- 30-cm (1-ft) section of dialysis tubing
- 50 mL tincture of iodine purchased or made as described in Getting Ready
- 10 drops starch solution

Extension for Part 2

- 1–2 drops of chicken liver blood
- 1–2 drops of a pure food color

Variation for Part 3

➤

- thin plastic sandwich bag

  Hefty® "Small Storage and Sandwich Bags" work well.

- plastic freezer bag
- 2 wide-mouthed containers
- 30 mL tincture of iodine
- 30 mL starch solution

Resources

The Sephadex resin can be purchased from Sigma Chemical Company, P.O. Box 14508, St. Louis, MO 63178, 800/325-3010.

- Sephadex—catalog # G-50-80 for 10 g

Pasteur pipets, Pasteur pipet bulbs, and rolls of dialysis tubing can be purchased from a chemical supply company such as Fisher Scientific, 9403 Kenwood Road, Suite C-208, Cincinnati, OH 45242, 800/766-7000.

- Pasteur pipets—catalog # 13-678-6A for 200 pipets
- Pasteur pipet bulbs—catalog # 14-065B for 72 pipet bulbs
- dialysis tubing—catalog # 08-667B for 100 feet

Tincture of iodine can be purchased at a pharmacy or made from sodium iodate, potassium iodide, and 3 M sulfuric acid. These can be purchased from a chemical supply company such as Flinn Scientific, P.O. Box 219, Batavia, IL 60510-0219, 800/452-1261.

- sodium iodate—catalog # S0218 for 25 g
- potassium iodide—catalog # P0066 for 25 g
- 3 M sulfuric acid—catalog # S0417 for 500 mL

Safety and Disposal

The Sephadex can be reused by storing it submerged in a saturated sodium chloride (NaCl) solution (40 g NaCl in 100 mL water). If a non-food refrigerator is available, use it to store the Sephadex—this will help to inhibit bacterial growth. Do not freeze the resin as this would damage the Sephadex beads. Sephadex can still be used if bacterial growth occurs. Remove the contaminated beads and place them in the trash. Use the remaining resin.

Goggles should be worn when working with the tincture of iodine. Iodine solutions stain some materials and may temporarily stain the skin. Students should be told to use caution while handling this material. If you choose to make your own tincture of iodine, you must exercise caution when working with sulfuric acid (H_2SO_4).

All excess chemicals in this experiment can be discarded by flushing down the drain with large amounts of water.

Getting Ready

Liquid laundry starch may be used for the starch solution in this activity. If laundry starch is not available, a saturated solution of starch may be made by adding 2 Tbsp corn starch to 500 mL cold water. Stir the solution vigorously and allow it to sit for 10–15 minutes. After the excess cornstarch has settled to the bottom of the container, pour the clear solution into labeled, plastic dropper bottles.

 If preparing tincture of iodine, use caution when working with sulfuric acid. See Safety and Disposal.

If tincture of iodine is not purchased, 500 mL of solution can be made by dissolving 0.23 g sodium iodate ($NaIO_3$) in 125 mL water in a 1-L beaker. To the sodium iodate solution add 5 g potassium iodide (KI). Mix thoroughly until dissolved. To the sodium iodate/potassium iodide solution add 30 mL 3 M sulfuric acid (H_2SO_4). Add water to the resulting solution to bring the final volume to 500 mL. Place the tincture of iodine in a dark bottle for storage. For the lab activity, pour some of the tincture of iodine into labeled dropper bottles. If you are doing Part 3 as a lab, each group will need about 50 mL tincture of iodine to cover the dialysis tubing.

To prepare a slurry of Sephadex resin: Add 1 g dry Sephadex resin to 100 mL water in a beaker. One gram dry Sephadex resin makes about 9–11 mL hydrated resin, which is enough to pack 8–10 pipet columns. Allow the resin to sit for 1 hour. After 1 hour, pour off the excess water and add fresh water so that there is at least a 2-in layer of water on top of the resin. Allow the resin to sit for another hour. Pour off any remaining water before dispensing the resin to the students.

To prepare the dialysis tubing: Soak the tubing in enough water to cover it for half an hour.

Opening Strategy

Demonstrate simple filtration by placing all the Styrofoam balls on top of the wire screen. Lift the screen. This will separate the two larger-sized balls from the smaller balls which pass through the screen as shown in Figure 1a.

Model the gel filtration by placing a piece of cardboard under the wire screen and arranging all the Styrofoam balls along the 1-ft side of the screen. Lift this end of the screen along with the cardboard backing as shown in Figure 1b. As the unit is lifted, the balls will roll down the screen with the largest balls rolling down first, followed by the medium-sized balls, and lastly by the smallest balls. The balls will separate by size because the largest balls spend less time within the openings of the screen while the smallest balls spend more time within the screen openings and therefore are retained longer as shown in Figure 1b. (You may wish to bend the last 3 inches of the wire screen upward to catch the balls as they reach the bottom of the incline.)

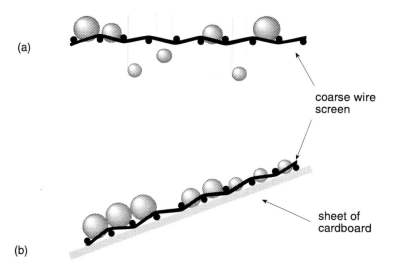

(a)

coarse wire
screen

sheet of
cardboard

(b)

Figure 1(a) Simulation of simple filtration
(b) Simulation of gel filtration

To familiarize the students with the starch indicator used in this activity, show them how the iodine and starch form the blue-black complex. This complex formation is used to indicate that the components have exited the column.

Procedure

Part 1: Preparing the Sephadex Column

1. Prepare a column by *loosely* blocking the bottom opening of the disposable Pasteur pipet with a small piece of crumpled tissue or filter paper. The tissue or filter paper can be pushed into place with a small piece of wire.

    ➤ **This paper will serve as a bed upon which the Sephadex resin (the column packing) will rest. Be aware that if the paper is packed too tightly in the bottom of the pipet, the liquids will not flow through the column.**

2. Using a disposable plastic pipet, add about 3 mL of the prepared Sephadex resin slurry to the Pasteur pipet so that the column is between ½–⅔ full of resin as shown in Figure 2.

    ➤ **If the column is too full, remove some of the excess resin with the pipet used to deliver the slurry. Stand the column upright in an empty cup or beaker until you are ready to use it.**

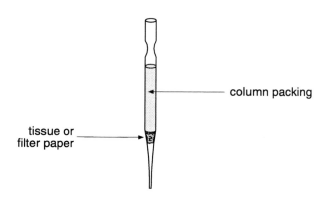

column packing

tissue or
filter paper

Figure 2: The packed column for gel filtration

Part 2: Gel Filtration

General Setup Information

1. For each trial, have someone hold the test tube upright, or place it in a test tube rack. Hold the narrow end of the column over the mouth of the test tube. The test tube will serve to collect the solution that exits the column.

2. Follow the specific procedures for each trial including addition of samples and preparation of the test tube and column. (See Figure 3.) Time or count and record the number of drops exiting the column starting with the first drop that appears after the addition of the first liquid and ending when the color appears or changes in the test tube. Observe the color of the band as it moves down the column as well as the color change in the test tube once the solution reaches the end of the column. Look for the same blue-black color of the starch-iodine complex as seen in Opening Strategy.

3. Between each trial, rinse the column by adding water drop-by-drop until the colored band elutes (drips off the column).

➤ **At no time should more than a couple of drops of water remain on the top of the column.**

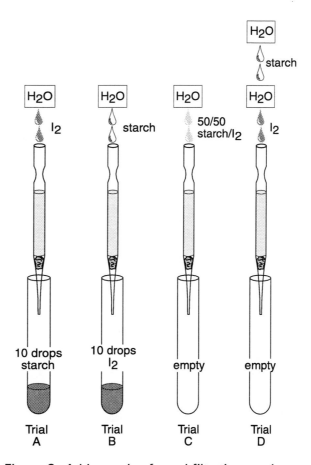

Figure 3: Add samples for gel filtration as shown.
Note that the same column is used for all trials.

Trial A: Iodine in Column

4. Add 10 drops of laundry starch or corn starch solution to a clean test tube.

5. Add 1–2 drops of tincture of iodine to the top of the packed Sephadex column prepared in Part 1. Add water drop-by-drop to the top of the column, counting the drops of liquid exiting, or use a timing device to record the amount of time. Stop counting or timing when the color in the test tube changes.

6. Record the number of drops or amount of time. Rinse the column with water.

Trial B: Starch in Column

7. Add 10 drops of tincture of iodine to a clean test tube.

8. Add 1–2 drops of starch solution to the top of the column. Add water drop-by-drop to the column, counting the drops of liquid exiting, or use a timing device to record the amount of time. Stop counting or timing when the color in the test tube changes.

9. Record the number of drops or amount of time. Rinse the column with water.

Trial C: Starch-Iodine Mixture in Column

10. Prepare a 50/50 starch-iodine mixture by adding 10 drops of starch solution and 10 drops of tincture of iodine to another clean test tube. Label the test tube 50/50 starch/iodine.

11. Place an empty, clean test tube under the column.

12. Add 1–2 drops of the 50/50 starch-iodine mixture to the top of the column. Add water drop-by-drop to the column, counting the drops of liquid exiting, or use a timing device to record the amount of time. Stop counting or timing when the first blue-black drop falls into the test tube.

13. Record the number of drops or the amount of time. Rinse the column with water.

Trial D: Iodine Followed by Starch

14. Place an empty, clean test tube under the column.

15. Add 1–2 drops of tincture of iodine to the top of the column and add water drop-by-drop. When the brown/yellow band of iodine reaches about halfway down the column, add 1–2 drops of starch solution. Add water drop-by-drop to the column, counting the drops of liquid exiting, or use a timing device to record the amount of time. Stop counting or timing when the color in the test tube changes.

16. Record the number of drops or amount of time. Rinse the column with water.

Part 3: Dialysis (Demonstration or Class Experiment)

1. Tie a knot at one end of the dialysis tubing that was soaked in Getting Ready.

2. Add 10 drops of starch solution to the inside of the dialysis tubing and close the other end with another knot.

3. Rinse the outside of the tubing with water and place it in a disposable cup.

4. Add enough of the tincture of iodine (approximately 50 mL) to the cup containing the dialysis tubing to just cover the tubing.

5. Observe the changes that occur in the cup/tubing over the next 10–20 minutes.

Extension for Gel Filtration

Compare chicken liver blood protein molecular size to food color molecular size. Repeat Part 2, Trials A and B, substituting blood for the starch and food color for the iodine. (Blood protein is a collection of macromolecules.)

Variation for Dialysis

Compare the dialysis ability of plastic bags by replacing the dialysis tubing with thin-walled plastic sandwich bags and thicker freezer bags. Place 15 mL (1 Tbsp) starch solution in each of the plastic bags and tie off or seal the bags. Stand each bag in a wide-mouthed container. Prepare a diluted tincture of iodine by mixing 30 mL (2 Tbsp) tincture of iodine with 60 mL (4 Tbsp) water.

Pour the diluted tincture of iodine into the wide-mouthed containers. The liquid in the container should cover the liquid in the bags. Prepare additional diluted tincture of iodine to use if needed. Let the bags set for several hours, making observations every half hour if possible. Since the tincture of iodine is dark brown-yellow, lift the bag out of the tincture of iodine to look for the color change. Watch the thin-walled plastic bag closely for the first half hour. A color change typically occurs in the first few minutes. Gradual darkening and spreading of the color will occur over the next 20–30 minutes. (Depending on the brand of sandwich bag you use, the length of time for color change may vary.)

 Most freezer bags will allow diffusion within 3 hours, but some bags require soaking overnight before any iodine will diffuse into the bag. The solution in the sandwich bag should change before the one in the freezer bag because of the difference in thickness of the plastic.

Discussion

- Ask students to decide from their data which compound (starch, iodine, or starch-iodine complex) exited the column first, second, and third; in other words, what was the order of elution?
  The starch-iodine complex exited first, the starch second, and the iodine last.

- Ask students what the order of elution tells you about the size of the molecules and their interactions with the Sephadex resin.
  Sephadex is a porous resin and small molecules can become trapped in the pores of the resin. If a molecule is too big to fit into the pores of the resin, it continues to travel down the column unimpeded. Thus, the larger the molecule, the less interaction with the Sephadex resin, and the faster it travels through the column. The starch-iodine complex is the largest molecule and the first to come out of the column. Uncomplexed starch is the next largest molecule; it separates out second. Iodine is composed of small molecules that get trapped in the pores of the resin and are the last molecules to come off the column.

- Discuss the changes that students observed in Part 3 as the tubing or bags sat in the tincture of iodine and the possible causes of the changes.
  The solution inside the tubing or bag takes on the dark blue-black color of the iodine-starch complex due to the diffusion of the iodine through the microscopic pores in the tubing or bags. The dialysis tube and the sandwich bag are very porous and allow molecules of a certain size to pass in and out through the pores. Since the starch molecules are so large, they will not be able to travel through the bag or tubing. Iodine

molecules are small enough to pass into the tubing or sandwich bag through the pores. Once inside, the iodine reacts with the starch to form the blue-black complex. Little noticeable change will occur outside the bags.

Explanation

Molecules can exist in different sizes ranging from as small as two atoms to as large as hundreds of thousands of atoms. Mixtures have a variety of sizes of molecules in them. Chromatography is a technique that can be used to separate a mixture into its individual components. A series of techniques classified as size-exclusion chromatography can be used to accomplish this separation based on the size of the molecules.

The most common technique used in size-exclusion chromatography is column chromatography or gel filtration. In gel filtration, the stationary phase or matrix consists of a column of porous beads whose pores are, on an average, comparable in size to the molecules to be separated. Larger molecules fit into fewer of these pores than do smaller molecules. As a result, the larger molecules spend less time on the stationary phase (within the pores of the gel) and more time in the mobile phase (the solution moving past the pores). A very large molecule may not fit into any of the stationary phase pores and will emerge from the column rapidly. Molecules of intermediate size will spend a small amount of time in the pores and will emerge later than the larger molecules. Small molecules will spend a relatively large amount of time in the pores and will be retained on the column for a considerably longer period of time. (See Figure 1b.)

Using the principles of gel filtration, a mixture of molecules of various sizes can be separated on a column. In Part 2 of this activity, each trial involves the formation of a complex between a macromolecule (starch) and a small molecule (iodine). The time it takes for each to emerge from the column gives us clues about their relative sizes. The blue-black starch-iodine complex is the largest molecule. The starch molecule is the next largest, and the iodine molecule is smallest. The starch-iodine complex molecules will emerge fastest and the iodine molecules slowest. See Table 1 on the following page for further explanation of Part 2.

Dialysis is another technique that separates molecules by size. Dialysis uses a semipermeable membrane to accomplish this task. The porous nature of a semipermeable membrane allows only molecules smaller than a certain size to pass through it. Molecules too large to fit through the pores of the membrane are not permeable. In this activity, the smaller iodine molecules moved through the dialysis membrane into the tubing or bags. The blue color that resulted in the tubing or the bags is the product of the reaction between iodine and starch. (See Figure 4.)

$$iodine\ (I_2)\ +\ starch\ \longrightarrow\ starch\text{-}iodine\ complex$$

(red-brown)   (colorless)   (blue-black)

Figure 4: Reaction between iodine and starch

Our kidneys extract the wastes from our blood using a dialysis process. Those molecules that are small enough to pass into the kidney are excreted. Those too large to pass into the kidney remain in the blood and are eliminated by other means.

Table 1: Observations and Explanations of Each Trial in Part 2

Trial	Solution Added to Column	Typical Results	Explanation of Observations
Part 2 A	iodine	Blue-black color appears in test tube. Time for iodine to travel through column: 310 s Number of drops: 80	Starch-iodine complex formed in test tube; small size of iodine molecule causes it to be held in column longer.
Part 2 B	starch	Blue-black color appears in test tube. Time for starch to travel through column: 105 s Number of drops: 30	Starch-iodine complex formed in test tube; starch molecule is larger than iodine molecule and will travel through the column faster.
Part 2 C	50/50 starch-iodine	Blue-black color travels down column. Time for complex to travel through column: 90 s Number of drops: 22	Starch-iodine complex is a larger complex than starch or iodine and will travel through the column fastest.
Part 2 D	iodine followed by starch	Blue-black color appears on column. Time for starch to travel through column: 40 s Number of drops: 10 Time for iodine to travel through column: 300 s Number of drops: 80	Starch molecules "catch up" to iodine molecules and form the starch-iodine complex on the column.

Key Science Concepts

- chromatography
- dialysis
- separation techniques

Cross-Curricular Integration

Language Arts
Use this activity with a mystery story in which separation and identification are important.

References

Boyer, R.F. *Modern Experimental Biochemistry;* Addison-Wesley: 1986; pp 87–96.

McLoughlin, D.J. "Size-Exclusion Chromatography: Separating Large Molecules from Small," *Journal of Chemical Education.* 1992, 69(12), 993–995.

Mewaldt, W.; Rodolph, D.; Sady, M. "An Inexpensive and Quick Method for Demonstrating Column Chromatography of Plant Pigments of Spinach Extract," *Journal of Chemical Education.* 1985, 62(6), 530–531.

Stryer, L. *Biochemistry,* 3rd ed.; W.H. Freeman: New York, 1988; pp 47–48.

Components of an Aspirin Tablet

What percentage of an aspirin tablet is really aspirin? You might be surprised by the answer. Aspirin tablets contain more than just the active ingredient acetylsalicylic acid. In this activity, students use various chemical techniques to examine common aspirin tablets for chemicals other than aspirin and determine how these chemicals affect the acidity and solubility of the tablet. They use this information to make an informed consumer decision about which brand of aspirin they would purchase.

Recommended Grade Level 7–12
Group Size .... 1–4 students
Time for Preparation .... 10–30 minutes
Time for Procedure .... 45–60 minutes

Materials

Procedure

Per Group
- 4 tablets of at least 3 of the following types of aspirin:
  - expensive brand (e.g., Bayer®)
  - inexpensive brand (e.g., Kroger Cost Cutter)
  - extra-strength brand (e.g., Extra Strength Anacin®)
  - "buffered" brand (e.g., Bufferin®)
  - enteric brand (e.g., Ecotrin®)
  - children's brand (e.g., St. Joseph's®)

 One of the brands of aspirin chosen for the experiment should be buffered. Keep the price labels on the products.

Procedure, Part 1

Per Group
- 3 or more small plastic bags (1 for each type of aspirin tested)
- (optional) hammer
- 2 pieces of blank typing paper
- 8½-in x 11-in (approximately) sheet of clear plastic wrap
- ½ tsp salicylic acid (available as Compound W wart remover)
- ½ tsp corn starch
- ½ tsp sugar (sucrose)
- ½ tsp talcum powder
- about 20 drops of tincture of iodine purchased or made as described in Getting Ready
- about 20 drops of 2% ferric chloride solution ($FeCl_3$) (See Getting Ready.)
- 2 dropper bottles or droppers
- paper towel
- glass stirring rod
- (optional) 1-L beaker
- (optional) dark bottle
- goggles

Procedure, Part 2

Per Group

- 3 pieces of universal pH paper about 2.5 cm (1 in) long
- 3 small test tubes or small glass containers
- test tube rack
- water
- 10-mL graduated cylinder
- 3–5 small plastic bags
- hammer
- 3 plastic bags
- masking tape for labels
- goggles

Procedure, Part 3

Per Group

- small test tubes or small glass containers
- 6–10 mL of 1 of the following acids:
  - 1 M hydrochloric acid (HCl) purchased or made from 40 mL 12 M hydrochloric acid and water
  - white vinegar (acetic acid solution, $HC_2H_3O_2$)

 Do not use red vinegar.

- 13.5 g baking soda ($NaHCO_3$)
- timing device with a second hand
- masking tape for labels
- (optional) 600-mL beaker

Variation

- non-aspirin products such as acetaminophen and ibuprofen

Extension

- access to a microwave

Resources

Salicylic acid, ferric chloride, hydrochloric acid, and pH paper can be purchased from a chemical supply company such as Flinn Scientific, P.O. Box 219, Batavia, IL 60510-0219, 800/452-1261.

- salicylic acid—catalog # S0341 for 50 g (can also be purchased as Compound W at a grocery store or pharmacy)
- sodium iodate—catalog # S0218 for 25 g
- potassium iodide—catalog # P0066 for 25 g
- 3 M sulfuric acid—catalog # S0417 for 500 mL
- ferric chloride—catalog # F0006 for 100 g
- 1 M hydrochloric acid (HCl)—catalog # H0013 for 500 mL
- concentrated hydrochloric acid (12 M HCl)—catalog # H0004 for 473 mL (1 pint)
- universal pH paper—catalog # AP1107 for package of 10 vials (100 strips per vial)

Aspirin products, tincture of iodine, salicylic acid (Compound W), corn starch, sugar, talcum powder, baking soda, and white vinegar can be purchased at a grocery store or pharmacy.

Safety and Disposal

Goggles should be worn when performing Parts 1 and 2 of this activity.

High doses of aspirin are toxic. Students should never taste any substance they prepare or use in the laboratory.

Iodine solutions can stain some materials and may temporarily stain the skin. Students should be warned to use caution when handling these materials. If you choose to make your own tincture of iodine, you must exercise caution when working with 3 M sulfuric acid.

If you choose to use a concentrated solution of hydrochloric acid to prepare the dilution needed for this activity, extreme caution should be used. Concentrated solutions of hydrochloric acid are very corrosive. They can cause severe chemical burns. The vapor is extremely irritating to the skin, eyes, and respiratory system. Work involving concentrated hydrochloric acid should be performed in a fume hood. Should contact occur, rinse the affected area with water for 15 minutes. If the contact involves the eyes, medical attention should be sought while the rinsing is occurring. When diluting concentrated acids, **always add acid to water** and not the reverse. The heat released in the dissolving process can cause splattering if the diluting is not carried out in the correct sequence.

Unused reagents can be diluted with water and flushed down the drain or saved for future use. At the end of the class period the students should roll up their "testing papers" and dispose of them in the wastebasket.

Getting Ready

Use caution when working with 3 M sulfuric acid. See Safety and Disposal.

If tincture of iodine is not purchased, 500 mL of solution can be made by dissolving 0.23 g sodium iodate ($NaIO_3$) in 125 mL water in a 1-L beaker. To the sodium iodate solution add 5 g potassium iodide (KI). Mix thoroughly until dissolved. To the sodium iodate/potassium iodide solution add 30 mL 3 M sulfuric acid (H_2SO_4). Add water to the resulting solution to bring the final volume to 500 mL. (In this situation, it is safe to add water to the acid solution because the solution is so dilute.) Place the tincture of iodine in a dark bottle for storage. For the lab activity, pour some of the tincture of iodine into labeled dropper bottles.

To prepare the approximately 2% ferric chloride solution, add 2 g ferric chloride ($FeCl_3$) to 100 mL water. Stir until the solid is dissolved. Pour into labeled dropper bottles.

See Safety and Disposal for information on handling hydrochloric acid.

If the 1 M hydrochloric acid (HCl) is not purchased, 500 mL of solution can be made by cautiously stirring 40 mL of 12 M hydrochloric acid (HCl) into 250 mL water in a 600-mL beaker. **A great deal of heat is evolved in this process.** After the solution has cooled to room temperature, dilute to a volume of 500 mL. Again, this acid solution is dilute enough that water can be safely added.

To prepare the baking soda solution ($NaHCO_3$), add 13.5 g baking soda ($NaHCO_3$) to 240 mL water.

Opening Strategy

Ask students what percentage of an aspirin tablet they think is really aspirin. Calculate this percentage as follows: Determine the mass of 10 buffered aspirin tablets in milligrams or grams and then calculate the mass of one tablet by dividing by 10. Check the box or bottle of aspirin for the aspirin content of each tablet recorded in milligrams. If necessary, convert the recorded aspirin content to grams to compare with the mass of the 10 tablets. Calculate the percentage of aspirin in each tablet by dividing the mass of aspirin per tablet recorded on the bottle by the mass of one aspirin tablet and then multiplying by 100. The resulting percentage should be less than 100%. Ask students for ideas on what makes up the remaining percentage of the tablet. Record the predictions on the board.

Procedure

Part 1: Examine the Composition of Aspirin Tablets

1. On a piece of typing paper, draw a box for each sample. Label the boxes: "corn starch," "sugar," "talcum powder," "salicylic acid," and the names of the different commercial aspirin products. (The corn starch, sugar, and talcum powder represent inactive aspirin ingredients.)

2. For each aspirin sample, use a hammer to crush two tablets in a small plastic bag (or crush the tablets with the heel of your shoe, protecting the aspirin in a bag wrapped in a paper towel).

3. Place the piece of plastic wrap on top of the labeled paper. Make two piles (about ¼ tsp each) of each substance in the appropriate box.

4. Place 1–2 drops of tincture of iodine on **one** of the piles for each substance and observe the reaction, if any. Record any observations, especially color changes.

5. Place 1–2 drops of 2% ferric chloride solution ($FeCl_3$) on the unused pile of each substance and observe the reaction, if any. Record any observations, especially color changes.

Part 2: Compare the pH Between Different Brands of Aspirin Products

1. Crush a tablet of one brand of aspirin product as described in Part 1, Step 2. Add the crushed tablet to 8 mL water in an appropriately labeled test tube and mix well by swirling.

2. Repeat Step 1 for each brand of aspirin product to be tested.

3. Test each aspirin solution with universal pH paper and record the pH for each brand of aspirin. Note which brand has the lowest pH and which brand has the highest pH.

4. Discard the solutions in the sink and flush them down the drain with large amounts of water.

Part 3: Compare the Time Required to Dissolve Aspirin Tablets in Acid

1. Label a clean test tube for each aspirin sample to be tested.

2. Add 3 mL 1 M hydrochloric acid (HCl) or vinegar (acetic acid, $HC_2H_3O_2$) to each test tube.

3. Place a tablet of one of the aspirin products into the appropriate test tube.

4. Without stirring, monitor and record the time (in seconds) that it takes for **most** of the tablet to dissolve.

➤ **Not all of the components of the aspirin tablet will dissolve. There will be a white, flaky solid remaining.**

5. Repeat Steps 2–4 for the remaining aspirin products.

6. Repeat this procedure in clean test tubes, substituting water for the acid, and then substituting baking soda solution.

7. Determine which aspirin product takes the longest time to dissolve and which takes the shortest time to dissolve in each liquid. Check the labels to see how the ingredients differ.

8. Discard the solutions in the sink and flush well with water.

Variation

- Perform this activity using non-aspirin products such as acetaminophen or ibuprofen instead of aspirin.

Extension

- Accelerate the aging process of aspirin by wrapping some aspirin tablets in a slightly damp paper towel and microwaving them for about 30 seconds. Use ferric chloride ($FeCl_3$) to test this "old" aspirin for the presence of residual salicylic acid.

Discussion

- Ask students what impurity was indicated by the color change for the substances tested with ferric chloride. What does a color change suggest about the presence of an impurity in the aspirin product tested?
  The color change from orange/yellow to purple indicates the presence of residual salicylic acid. The salicylic acid is present from unreacted starting material as well as from the hydrolysis that occurs as the aspirin ages. The cotton in the bottle soaks up moisture and slows down the hydrolysis. The residual salicylic acid is present in all of the aspirin products tested. The "old" aspirin tested in the Extension will give a deeper purple color because it contains more residual salicylic acid due to hydrolysis.

- Ask students what was indicated by the color change in the substances tested with iodine. Have them try to explain the reason for the presence of starch in the aspirin product.
  The color change from light brown to blue-black indicates the presence of starch. Starch is used in an aspirin tablet as a binder to hold the tablet together.

- Have students explain the significance of the data gathered in Part 3.
  In Part 3, the aspirin products are timed to see how fast they would dissolve in an acid, water, and a base. The acid simulates what happens in stomach acid, and the base simulates what happens in the intestine. If the tablet dissolves quickly in acid, it would be absorbed quickly through the stomach into the blood stream to the source of pain. If the tablet dissolves in the base but not in the acid, then the aspirin would be absorbed in the intestine.

- As a class, list the ingredients in aspirin tablets based on the tests just performed. Ask the students to figure out what the major non-active ingredient is by examining the labels of several aspirin bottles. If Bayer aspirin was used in the activity, it would be appropriate to ask the students how true Bayer's claim of "100% aspirin" is.
  Aspirin tablets contain aspirin, starch (several types may be listed on the package), talc, wax, and residual salicylic acid. The major non-active ingredient is the starch, which is

used as a binder. *Bayer's claim that their aspirin is "100%" aspirin is not necessarily true because of the presence of the starch and residual salicylic acid.*

- List other types of analgesic products available besides aspirin. Discuss reasons why these products were developed if aspirin works so well. Who should use these alternative products? *Ibuprofen and acetaminophen were developed to replace aspirin for people who were allergic to aspirin or those whose stomachs could not tolerate aspirin or buffered aspirin.*

- Have students choose which aspirin product they would purchase if they had a headache based on the data that they gathered in this activity.
  The students should consider acidity, dissolving time, and cost when making their decisions. Students may also rely on advertising claims and past experience. Students could make a bar graph that compares cost/active ingredient for all brands of aspirin.

Sample Data

substance tested	reaction with iodine	reaction with FeCl$_3$	time for dissolving		
			in acetic acid	in water	in NaHCO$_3$ solution
Norwich Max	black (reaction)	lavender (reaction)	.83 min	1 min	2.5 min
Bufferin	black (reaction)	lavender (reaction)	2 min	2 min	2 min
Meijer Buffered	black (reaction)	lavender (reaction)	6 min	6 min	6 min
Bayer	black (reaction)	lavender (reaction)	1 min	1 min	6.5 min
Anacin	black (reaction)	lavender (reaction)	1.5 min	1.5 min	7 min
Kroger	black (reaction)	lavender (reaction)	2 min	2.5 min	17 min
Generic	black (reaction)	lavender (reaction)	5.5 min	5.5 min	1 min
Ecotrin*	tan (no reaction)	lavender (reaction)	**	**	20 min
starch	black (reaction)	(orange (no reaction)			
sugar	black (reaction)	orange (no reaction)			
talc	tan (no reaction)	orange (no reaction)			
salicylic acid	tan (no reaction)	purple (reaction)			

* Ecotrin has a heavy coating that retards dissolving in water or acid; once the coating has dissolved away, the tablet dissolves quickly in baking soda (about 1min).

** No significant dissolving occurs within three days.

Explanation

Aspirin tablets contain more than just aspirin. In addition to the acetylsalicylic acid, they contain binders used to hold the tablet together. If starch (a typical binder) is present, testing with iodine will produce a bluish-black color. A purple color upon adding ferric salts (Fe^{3+})

indicates the presence of salicylic acid, one of the starting materials used to make aspirin. Tablets may also contain coatings and/or flavorings.

The pH is a measure of the acidity or alkalinity of a substance. Materials that fall below seven on the pH scale are classified as acids. Those that are above seven are bases. The lower the pH number, the more acidic a substance is. Aspirin will fall below seven because it contains acetylsalicylic acid. In addition, any excess salicylic acid and acetic acid remaining from the production of the aspirin will also contribute to the acidity of an aspirin tablet. Although purification steps are performed, some small amount of starting materials (salicylic acid and acetic acid) remain in the aspirin. (See Activity 10, "Making Aspirin and Oil of Wintergreen," for a more detailed description of the synthesis of aspirin).

In the acidic conditions of the stomach, uncoated aspirin tablets typically dissolve quickly and are absorbed by the bloodstream. Certain aspirin products have coatings which are resistant to dissolving in the acid environment of the stomach. These aspirins are called enteric. Part 3 shows that the coating on enteric aspirin does dissolve in a base such as baking soda solution, which is used to simulate the basic nature of the small intestine.

Key Science Concepts

- acids and bases
- buffers
- indicators

Cross-Curricular Integration

Home, Safety, and Career
Discuss truth in advertising, the use of advertisements to sell products, and the use of product labels. This should help students to make more informed consumer decisions.

Language Arts
Have students create advertising slogans based on data from the activity.

References

American Chemical Society. *Chemistry in the Community;* Kendall/Hunt: Dubuque, Iowa, 1988; p 442.

Comets Science Volume II: Career Oriented Modules to Explore Topics in Science; Department of Curriculum and Instruction, School of Education, University of Kansas. NSTA, 1982; 71.

Hill, J.W. *Chemistry for Changing Times;* Macmillan: New York, 1988; pp 486–491.

"Is Bayer Better?" *Consumer Reports.* July 1982, 347–349.

"The New Pain Relievers," *Consumer Reports.* November 1984, 636–638.

"Pills that Compete with Aspirin," *Consumer Reports.* August 1982, 397–399.

Tocci. *Chemistry Around You: Experiments and Projects with Everyday Products;* Arco: New York, 1985, pp 75–82.

Cost/Mass Analysis of Different Aspirin Brands

Aspirin is the world's most commonly used analgesic, or pain reliever. There are many brands of aspirin on the market, each advertised as being effective. If aspirin consists mainly of acetylsalicylic acid, how do the various brands differ? Are you paying too much for what you get? In this activity, students examine advertising claims for various brands of aspirin and determine the scientific basis (or lack thereof) for each claim.

Recommended Grade Level 4–12
Group Size .... 1–4 students
Time for Preparation .... 10 minutes
Time for Procedure .... 30–45 minutes

Materials

Procedure

Per Class
- 50–100 tablets of each of several brands of aspirin products, including the following:
  - expensive brand (e.g., Bayer®)
  - inexpensive brand (e.g., Kroger®)
  - extra-strength brand (e.g., Extra Strength Anacin® or Bayer)
  - "buffered" brand (e.g., Bufferin®)
  - children's brand (e.g., St. Joseph's®)

 Make sure that all the products are aspirin and not acetaminophen or ibuprofen. Also, keep the aspirins in the original bottles and make sure the price label and the number of tablets originally present appears somewhere on the package.

- balances calibrated in milligrams if possible (Calibration in grams will work.)
- calculators

Variation
- acetaminophen or ibuprofen products

Safety and Disposal

High doses of aspirin are toxic. Students should never taste substances that are prepared or used in the laboratory.

Samples should be reused for other activities if possible. Otherwise, the aspirin can be discarded in the wastebasket or dissolved in water and flushed down the drain.

Opening Strategy

Present the class with an extra-strength aspirin product and a regular-strength aspirin product. Discuss the advertising claims of each. (It may be helpful for students to bring in aspirin advertisements from magazines or discuss television commercials for aspirin.) Tell students that you have a "pounding headache." Based on the information gathered from the advertisements, have students choose the aspirin product you should take to relieve your headache and explain their choice. Have students consider the amount of aspirin in the product and cost. Record their selection on the board and then perform the activity.

Procedure

1. Record the price of each bottle of aspirin and the number of tablets in each bottle.

2. Determine the mass of 10 tablets of each brand of aspirin. Record the mass in milligrams or grams. Calculate the mass (in mg or g) of a single tablet by dividing the total weight by 10. Record the results and return the tablets to the bottle for use in other experiments.

3. Determine the mass (in mg or g) of aspirin in each tablet by examining the product labels. Record this information.

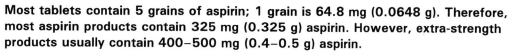

   Most tablets contain 5 grains of aspirin; 1 grain is 64.8 mg (0.0648 g). Therefore, most aspirin products contain 325 mg (0.325 g) aspirin. However, extra-strength products usually contain 400–500 mg (0.4–0.5 g) aspirin.

4. Calculate the percentage of aspirin in each tablet by dividing the mass of aspirin in each tablet by the mass of each tablet and multiplying by 100.

5. Calculate the cost of each tablet by dividing the price of the bottle by the number of tablets in the bottle.

6. Calculate the cost of 100 mg (0.1 g) of aspirin by dividing the cost per tablet by the milligrams or grams of aspirin in each tablet and multiplying by 100.

7. Construct a line graph or histogram of the cost per 100 mg (0.1 g) of aspirin versus the brand of aspirin.

Variation

• Conduct this activity using acetaminophen or ibuprofen products.

Discussion

• Ask students to write down television advertising claims for aspirin and bring them to class. Discuss the claims made about different brands with regard to the data that you collected during this activity. Ask the students if Bayer is really "100% pure aspirin" and if Anacin contains "50% more pain reliever than regular aspirin." Ask students if the products' claims are true or false.

Bayer cannot be 100% pure aspirin because there are binders and coatings present in every tablet. (See Activity 8, "Components of an Aspirin Tablet.") The class will have to decide whether they consider these claims to be deceptive or not. The claim by Anacin of "50% more aspirin than regular-strength tablets" can be considered true because regular-strength tablets contain 325 mg aspirin, 50% of which is 163 mg; 325 mg + 163 mg = 488 mg. An extra-strength Anacin or Bayer tablet contains 500 mg aspirin, or roughly 150% of the aspirin in a regular-strength tablet. Advertisements may not be false, but they can be misleading. Consumers are sometimes led to believe something about products that is really not true.

- Ask the students why they weighed 10 tablets instead of 1 or 100 tablets.
  Weighing only one small object can lead to significant sampling errors. Each tablet in a bottle of aspirin will have a slightly different mass. It is better to take an average of a sample of tablets to get a reasonable estimate for each tablet. Ten tablets was a convenient number, given the sensitivity of most balances. Although 100 tablets would give a better average, it is a large number and errors can occur in counting. Variation in individual sample weights (necessary information for quality control in industry) can be more easily detected if smaller sample sizes are used.

- Pose this question to the class: Is it more cost effective to take three regular-strength aspirin or two extra-strength aspirin? Have students calculate the cost for the two choices based on the data gathered in this activity. Discuss possible reasons why aspirin companies advertise the extra-strength brands as the better choice.
  Regular aspirin will provide more pain relief at a lower cost than the extra-strength aspirin. The extra-strength brands are marketed so that people will think that they provide more pain relief. Since the extra-strength aspirin tablets have "more medicine" than regular strength aspirin tablets, people will pay more for the extra-strength aspirin.

Sample Data

substance tested	mass of 10 tablets (mg)	mass of 1 tablet (mg)	mass of aspirin (mg)	% aspirin per tablet	price per number of tablets	cost per tablet	cost per 100 mg of aspirin
Generic	3730	373	325	87.1 %	$2.99/300	$0.0100	$0.00307
Anacin	5230	523	400	76.4 %	$4.49/50	$0.0898	$0.0225
Bayer	4060	406	325	80.0 %	$3.99/50	$0.0798	$0.0246
Bufferin	6730	673	325	48.3 %	$4.79/50	$0.0958	$0.0295
Extra Strength Bayer	6225	623	500	80.3 %	$4.59/50	$0.0918	$0.0184

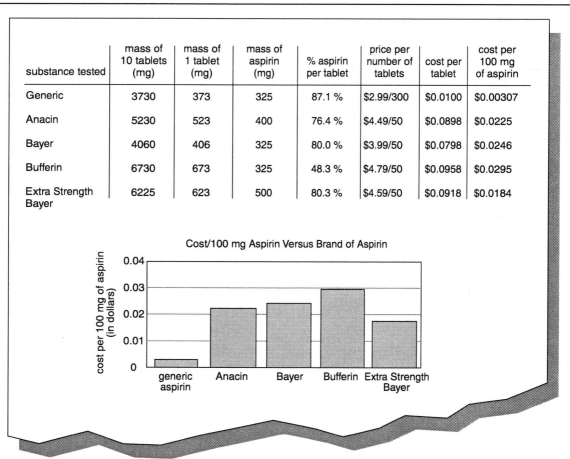

Cost/100 mg Aspirin Versus Brand of Aspirin

Explanation

In North America alone, over 55 billion tablets of aspirin are consumed each year. Nearly all of the aspirin sold, regardless of the brand name on the label, is made by two companies, Dow Chemical Company and Monsanto. Besides aspirin, Dow Chemical produces polymers. Monsanto produces fertilizers, insecticides, and herbicides in addition to aspirin.

It is a little-known fact that an aspirin tablet is not 100% acetylsalicylic acid but also has other ingredients. The other ingredients, like starch, are used in the preparation of the tablet. Other ingredients may be used for coatings, flavorings, or buffers (if specified).

Information is not presented in the same manner on all aspirin labels. Some brands report the amount of aspirin in each tablet in different units of measure. The mass of aspirin in each tablet can be expressed in grains, grams, or milligrams. One grain is equal to 64.8 mg and most tablets contain 5 grains. Some products report this as 325 mg. Extra-strength products usually contain between 400–500 mg of aspirin per tablet.

Key Science Concept

- cost/mass analysis

Cross-Curricular Integration

Home, Safety, and Career
Using the data gathered during the experiment, students can evaluate the different brands of aspirin as to the truth in advertising claims and consumer value.

Language Arts
Have students write to consumer groups and manufacturers asking for information that justifies their advertising claims regarding aspirin.

Study advertisements for aspirin and identify the rhetorical strategies used to persuade people to buy the products.

Develop a commercial for a brand of aspirin and perform it for the class.

Life Science
Discuss possible uses for aspirin besides pain relief (For example, it is prescribed for patients with heart problems) and side effects of aspirin. (For example, young children should not be given aspirin for fever because of the risk of Reyes Syndrome.)

Mathematics
Have students gather the data during science class, then conduct the calculations during math class, focusing on the percentage calculations, unit conversions (grams and grains to milligrams), graphing, and division skills.

Social Studies
Have students research the discovery of aspirin and its current production.

References

American Chemical Society. *Chemistry in the Community;* Kendall/Hunt: Dubuque, IA, 1988; p 442.

Hill, J.W. *Chemistry for Changing Times;* Macmillan: New York, 1988; pp 486–491.

"Is Bayer Better?" *Consumer Reports.* July 1982; 347–349.

"The New Pain Relievers," *Consumer Reports.* November 1984; 636–638.

"Pills that Compete with Aspirin," *Consumer Reports.* August 1982; 397–399.

Tocci, S. *Chemistry Around You: Experiments and Projects with Everyday Products;* Arco: New York, 1985; pp 75–82.

Making Aspirin
and Oil of Wintergreen

Have you ever smelled an aspirin? How about oil of wintergreen? Different? Most definitely. Despite these distinctly different odors, this activity shows that the two are made from salicylic acid.

Recommended Grade Level 9–12
Group Size .... 1–4 students
Time for Preparation .... 30 minutes
Time for Procedure .... 60 minutes

Materials

Procedure

Per Group
- 2 medium-sized test tubes
- test tube holder or spring clothespin
- 1 g salicylic acid
- 20 mL ice water
- disposable, plastic pipet OR filter paper and a funnel
- spatula
- thermometer to check temperature of warm- and hot-water baths
- goggles

Per Class
- 2–3 hot-water baths, each made from the following:
  - 250-mL beaker of water
  - hot plate OR Bunsen burner, ring stand, ring clamp, and wire gauze
- ice baths, each made from the following:
  - 250-mL beaker
  - ice water
- 100 mL 2% ferric chloride solution ($FeCl_3$) in a dropper bottle (See Getting Ready.)
- plastic gloves
- container in which to prepare 100 mL ferric chloride solution
- paper towel or filter paper
- 20 mL concentrated sulfuric acid (18 M H_2SO_4) in a glass dropper bottle
- 20 mL acetic anhydride in a dropper bottle
- 50 mL methyl alcohol (wood alcohol, CH_3OH)

 The instructor should dispense the sulfuric acid, the acetic anhydride, and the methyl alcohol.

Resources

The following items can be purchased from a chemical supply company like Flinn Scientific, P.O. Box 219, Batavia, IL 60510-0219, 800/452-1261.

- salicylic acid—catalog # S0001 for 100 g
- acetic anhydride—catalog # A0156 for 100 g
- methyl alcohol—catalog # M0054 for 500 mL
- sulfuric acid (18 M)—catalog # S0228 for 100 mL
- ferric chloride—catalog # F0006 for 100 g
- disposable, plastic pipets—catalog # AP1516 for 500 pipets

Methyl alcohol can also be purchased from a hardware store as a paint thinner.

Safety and Disposal

Goggles should be worn when performing this activity. Concentrated solutions of sulfuric acid and acetic anhydride are very corrosive. They can cause severe chemical burns. The vapor is extremely irritating to the skin and eyes. Should contact occur with either material, rinse the affected area with water for 15 minutes. If the contact involves the eyes, medical attention should be sought while the rinsing is occurring. The vapor from acetic anhydride is also particularly irritating to the respiratory system and should be used only in a fume hood or in a room with a very good exhaust system. Both the concentrated sulfuric acid and the acetic anhydride should be dispensed by the instructor. Wear gloves to protect your hands while dispensing either of these materials.

Methyl alcohol (wood alcohol, CH_3OH) is highly flammable and toxic. It can cause blindness or death through ingestion, absorption through the skin, or inhalation. Methyl alcohol should be dispensed by the instructor. Wear gloves to protect your hands while dispensing the alcohol.

Oil of wintergreen is toxic in large quantities. Avoid ingestion or contact. Wear gloves to protect your hands when handling oil of wintergreen. Should contact occur, rinse the affected area with water for 15 minutes. If the contact involves the eyes, medical attention should be sought while the rinsing is occurring.

Unused reagents can be diluted with water and flushed down the drain or saved for future use. The acetic anhydride and sulfuric acid must be tightly capped to prevent them from absorbing moisture from the air. Discard the crude aspirin product in the trash can and pour the oil of wintergreen down the drain with large amounts of water.

Getting Ready

To prepare an approximately 2% ferric chloride solution, add 2 g ferric chloride ($FeCl_3$) to 100 mL water. Stir until the solid is dissolved. Pour into a labeled dropper bottle.

To prepare a hot-water bath, fill a 250-mL beaker about half-full with water. Heat the water on a hot plate or over a Bunsen burner to approximately 70–80°C. To prepare a warm-water bath, fill a 250-mL beaker about half-full with hot tap water that is approximately 50°C. To prepare an ice bath, put cold water and ice in a 250-mL beaker. To maintain a temperature close to 0°C, make sure that the bath always contains some solid ice.

Opening Strategy

You may want to begin the lesson by discussing some of the uses and properties of aspirin. Talk about the pharmaceutical industry in general, and the processes whereby new drugs are discovered, developed, and put on the market. Discuss, for example, the fact that most drugs have been developed from chemical compounds originally found in plants. Discuss or have students find out on their own how aspirin was discovered. Show the class the structures of aspirin, salicylic acid, and oil of wintergreen and ask them what common features are shared by the structures of salicylic acid, acetylsalicylic acid (aspirin), and methyl salicylate (oil of wintergreen). Ask them if they know in which plant salicylic acid was discovered. (See the Explanation for details on the history of aspirin.)

Procedure

Part 1: Preparation of Acetylsalicylic Acid

1. Place 0.5 g salicylic acid into a dry, medium-sized test tube.

 The instructor should add the acetic anhydride in the next step. If contact occurs, rinse the affected area with water for 15 minutes. If contact involves the eyes, medical attention should be sought while rinsing is occurring.

2. Add approximately 1.0 mL (20 drops) acetic anhydride and mix by gently swirling the test tube.

 The instructor should add the sulfuric acid in the next step. If contact occurs, rinse the affected area with water for 15 minutes. If contact involves the eyes, medical attention should be sought while rinsing is occurring.

3. Carefully add one drop of concentrated sulfuric acid (18 M H_2SO_4) and swirl again.

4. Place the tube in a warm-water (about 50°C) bath for 15 minutes. Do not allow water to get into the test tube because it will interfere with the reaction.

 If the test tube is heated in near-boiling water, the heating time is reduced to approximately 5 minutes. However, caution must be exercised to keep boiling water from splattering into the test tube.

5. Remove the test tube with a test tube holder or spring clothespin and cool in an ice bath. Add about 2 mL (35–40 drops) of ice water (taken from the ice bath) to the test tube in the ice bath. Leave the test tube in the ice bath until crystals form.

6. Separate the crude acetylsalicylic acid (aspirin) crystals from the solvent by drawing off the remaining liquid with a pipet. (Alternatively, filter the mixture with a funnel and filter paper.)

7. Wash the crystals twice with a small amount of ice water, removing the excess water as in Step 6 each time.

8. Empty the solid from the test tube onto a piece of filter paper or paper towel to dry.

9. Test for residual salicylic acid by placing three drops of 2% ferric chloride solution ($FeCl_3$) on top of a small sample of the crude aspirin. Record any observations.

Part 2: Preparation of Methyl Salicylate

 Never smell a compound in the laboratory by holding it directly under your nose. For Steps 1 and 4, use the wafting technique described in Employing Appropriate Safety Procedures.

1. Place 0.5 g salicylic acid into a test tube and note its odor. Record any observations.

 The instructor should add the concentrated sulfuric acid (18 M H_2SO_4) and methyl alcohol (CH_3OH). Handle both solutions with extreme care. If contact occurs, rinse the affected area with water. If the contact involves the eyes, medical attention should be sought while rinsing is occurring.

2. Add 3 mL (60 drops) methyl alcohol and 1 drop of concentrated sulfuric acid. Mix by gently swirling the test tube.

3. Heat the mixture in a hot-water bath (70–80°C) for about 15 minutes.

4. Note the odor of the liquid that remains in the test tube. Identify the odor produced by this compound. Record any observations.

Discussion

- What was the function of the sulfuric acid (H_2SO_4)?
  The sulfuric acid serves as a catalyst that speeds up the reaction by extracting the alcohol group (–OH) from the salicylic acid.

- What are some similarities and differences between the two chemicals that were produced in this activity?
  Both are esters, are derivatives of salicylic acid, and have anti-inflammatory properties. However, methyl salicylate is a phenol and has a strong odor, while acetylsalicylic acid is a carboxylic acid with no strong odor.

Explanation

There are a variety of pharmaceutical agents available for relief from pain. These drugs are known as analgesics, the most common of which is aspirin.

Aspirin was first described in the science literature in the late 1700s when it was noted that mixtures of powdered willow bark relieved the pain symptoms caused by malaria. However, it wasn't until the late 1800s that the active compound, salicylic acid, was isolated from willow bark. Besides being a pain reliever, it was noted that salicylic acid also decreased fevers (an antipyretic) and reduced inflammation (an anti-inflammatory agent). The acidity of salicylic acid caused it to exhibit toxic effects in this purified form. Chemists attempted to modify the structure of salicylic acid to decrease its toxicity. The first attempt was to neutralize the acid by forming the sodium salt. Although the sodium salt did not irritate the mouth as much as the original salicylic acid did, it was still very irritating to the stomach. Further attempts to modify the structure eventually led to the esterification of the acid functional group. Several different esters were attempted before acetylsalicylic acid (more commonly called aspirin) was made. Not only was aspirin far less acidic than the original salicylic acid, it also retained the analgesic, antipyretic, and anti-inflammatory properties of the original compound. Aspirin is also now known to inhibit the clotting of blood (an anticoagulant).

In the reaction shown in Figure 1, the sulfuric acid (H_2SO_4) acts as a catalyst. This is why only a small amount of it was needed to perform the experiment. It is used to help remove the –OH group from the salicylic acid.

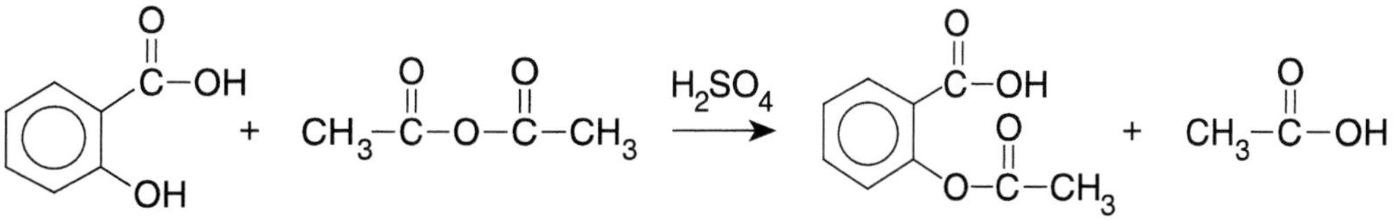

salicylic acid | acetic anhydride | aspirin | acetic acid

Figure 1: Synthesis of aspirin from salicylic acid

Methyl salicylate, which was synthesized in Part 2 (See Figure 2), has many of the same anti-inflammatory properties as aspirin. This compound is commonly known as oil of wintergreen and is one of the components of "witch hazel." Oil of wintergreen is often used in products like witch hazel or liniments that are intended to reduce skin inflammation or muscle aches.

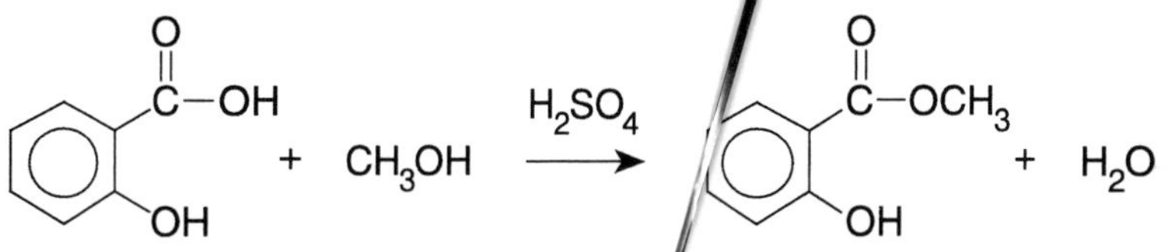

salicylic acid | methyl alcohol | methyl salicylate | water

Figure 2: Synthesis of methyl salicylate from salicylic acid

Methyl salicylate is also an ester. Esters are formed when a carboxylic acid reacts with an alcohol. The process is known as esterification. During esterification the –OH group on the acid combines with an H from the alcohol and H_2O is released. This reaction typically requires a catalyst to speed the reaction. Concentrated sulfuric acid was used as the catalyst due to its dehydrating capability. A sample equation for esterification is shown in Figure 3.

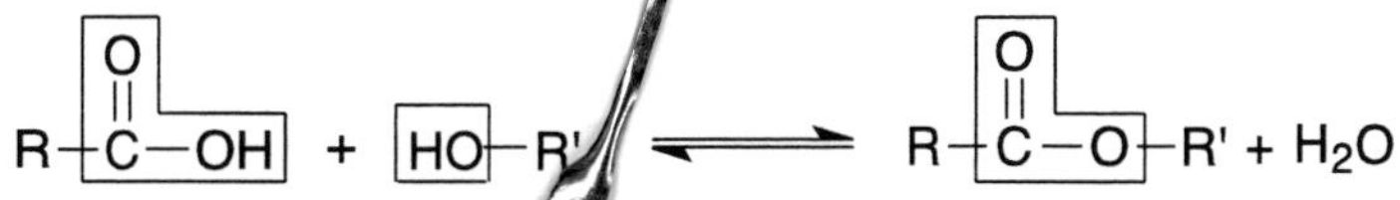

carboxylic acid | alcohol | ester

Figure 3: Reaction of a carboxylic acid and an alcohol to form an ester and water

When low-molecular-weight carboxylic acids are esterified, the resulting esters are typically colorless liquids with fruity odors. These synthetic esters are used in the food industry as artificial flavorings. In many cases the esters produced in the laboratory are the same molecules that give natural fruits their characteristic flavors. For example, isoamyl acetate, the chemical that gives bananas their characteristic flavor, can be made in the lab by reacting isoamyl alcohol with acetic acid. Other synthetic esters have no natural counterparts; however, they do have fruity flavors that can be used in foods.

It is interesting to see that by changing the functional groups of the basic salicylic acid molecule, compounds that are similar in some respects but very different in others can be created. By changing functional groups on a molecule, organic chemists can synthesize compounds that retain desirable properties with a fewer number of undesirable side effects.

Figure 4 shows four different molecules that can be made by changing the functional groups on salicylic acid.

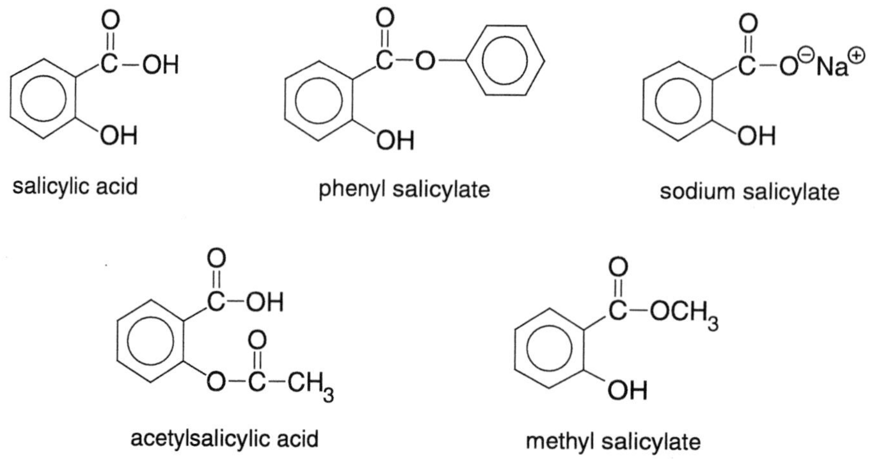

salicylic acid   phenyl salicylate   sodium salicylate

acetylsalicylic acid   methyl salicylate

Figure 4: The structures of salicylic acid and some of its derivatives that have been used as analgesics

Key Science Concepts

- acids and bases
- catalysts
- chemical changes
- chemical reactions
- functional groups

Cross-Curricular Integration

Life Science
Ask students how the fact that most drugs have been developed from chemical compounds originally found in plants is related to some current environmental problems regarding the loss of biological diversity in the rain forests.

Social Studies
Research the history of the pharmaceutical industry and the processes whereby new drugs are discovered.

Research the history of analgesics and their changes over the years.

References

Hill, J.W. *Chemistry for Changing Times*, 4th ed.; Burgess: Minneapolis, MN, 1984.

Jarvis, B., Mazzocchic, P. *Form and Function—An Organic Chemistry Module;* Interdisciplinary Approaches to Chemistry; Harper & Row: New York, NY, 1978; pp 71–72.

Tocci, S. *Chemistry Around You: Experiments and Projects with Everyday Products;* Arco: New York, NY, 1985; pp 75–82.